INTRODUCTION TO THE SCIENTIFIC STUDY OF
ATMOSPHERIC POLLUTION

INTRODUCTION
TO THE SCIENTIFIC STUDY OF
ATMOSPHERIC POLLUTION

Edited by

B. M. McCORMAC

*Lockheed Palo Alto Research Laboratory,
Palo Alto, Calif., U.S.A.*

D. REIDEL PUBLISHING COMPANY

DORDRECHT-HOLLAND

Library of Congress Catalog Card Number 70–170340

ISBN 978-90-277-0243-2   ISBN 978-94-010-3137-0 (eBook)
DOI 10.1007/978-94-010-3137-0

PREFACE

The Editor undertook the preparation of this book for two reasons. The first was to fulfill the need for an introductory level book combining the multidisciplinary aspects of atmospheric pollution. This book does cover most of the key facets: sources and sinks of atmospheric pollutants, atmospheric chemistry, transport and meteorology, effects on human beings and vegetation, and surveillance of air quality. The length of the book was purposely limited so that the cost would be within the means of most interested users.

The second reason for preparing this book was stimulated by my mother-in-law, Mrs Clayton D. Root, Councilwoman and member of the Planning Commission of Crown Point, Indiana, who pointed out that all governmental levels must make decisions affecting the atmospheric quality of their regions of responsibility and yet have little information on atmospheric pollution for their assistance. For the most part the achievement of air quality standards is a local problem and every locality is unique.

This book is written to serve the needs of the introductory text for students and researchers starting into atmospheric pollution research or studies. Most of the book is also readable by laymen who desire to learn about the subject and those who must fulfill some governmental or advisory responsibility concerning atmospheric pollution for society.

The authors of the various chapters have made excellent presentations in a concise and clear fashion and have provided extensive references and bibliographies for those desiring to go beyond this book. The authors and the publisher have made a special effort for rapid publication of an up-to-date status of atmospheric pollution which is an ever-changing research area.

The assistant editor, Mrs Diana Root McCormac, checked the manuscripts and proofs and worked hard to achieve a uniform style throughout the book.

The Editor is especially appreciative of the interest, assistance, and encouragement of the Lockheed Palo Alto Research Laboratory for the preparation of this book.

BILLY M. McCORMAC

Palo Alto, September 1971

TABLE OF CONTENTS

PREFACE   V

BILLY M. McCORMAC and ROBERT VARNEY / Introduction   1

ROBERT VARNEY and BILLY M. McCORMAC / Atmospheric Pollutants   8

R. W. SHAW and R. E. MUNN / Air Pollution Meteorology   53

RICHARD L. MASTERS / Air Pollution – Human Health Effects   97

SAMUEL N. LINZON / Effects of Air Pollutants on Vegetation   131

GEORGE B. MORGAN and GUNTIS OZOLINS / Air Quality Surveillance   152

GLOSSARY   164

INDEX OF SUBJECTS   167

INTRODUCTION

BILLY M. McCORMAC and ROBERT VARNEY

Lockheed Palo Alto Research Laboratory, Palo Alto, Calif., U.S.A.

1. Importance of Understanding Air Pollution

Air pollution is a far older problem than is sometimes recognized today. Some fifty years ago, a cartoon enjoyed wide publicity. It showed two New York City boys sitting on a fence in the country at a 'fresh air camp', one of them visibly sniffing at some strange odor. His companion was saying, "That's the fresh air you smell, but you'll get used to it."

The condition of Earth's air environment is very important for man and for all of the flora, fauna, and materials exposed to the air. Air pollution adversely affects our comfort and health. Economic effects result from reduced productivity, increased cleaning requirements, and decreased lifetime of materials. The natural beauty of areas can be degraded. Air pollutants affect the very existence of certain species in certain regions. Air pollution is receiving increased attention in all parts of the world. Although there has been some improvement in the detection of air pollutants, there has mainly been a change in man's acceptance of air pollution. Many of the types and sources of pollutants have been readily identified for years. Instead of putting up with the degrading effects of air pollution or moving, people are now demanding that the sources of air pollution be eliminated. Society, through its various organizations and structure, must and will make a continuing number of decisions affecting the amount of air pollution. Such decisions should be based on an understanding of the total system of air pollutants. The economic, social, and technical trade-offs for various courses of action should be apparent. Such an understanding requires the efforts of many diverse disciplines and research areas. The technical base of information is currently very meager and must be improved through research and development programs.

Pollution of Earth's air environment is an international problem, as the complex cycles and transport of air pollutants are unaffected by political boundaries. The combination of an increasing population and an improving standard of living will greatly increase the world's air pollution problems. The transit of air pollutants across international boundaries will probably occur more frequently in the future. Improvement involves both technical and political considerations. International agreements are necessary, but these must be based on a technical understanding of the overall cycle of the air pollutants.

The fundamental decisions to be made by society concerning air pollution are not scientific. Society must select a set of long range air standards considering economics, population, standard of living (energy requirements), recreation, health, safety, transportation, and political considerations. A scientific understanding of air pollutants, effects, cycles, etc., will aid society in making the best decisions to achieve the goals. Elimination of all air pollutants is impossible as many of them have natural sources. Therefore, some realistic compromises are necessary. Better technical knowledge will play a key role in the analyses and comparison of various courses of action.

More research on Earth's air environment is obviously required. It is unlikely that we will ever have completely adequate information about all facets of air pollution; however, society must take courses of action essentially on a daily basis. The best use of the available facts and predictions must be made both to influence future research requirements and to make decisions. There are commissions and decision making bodies from the towns and counties on up through the state and federal governments. Laws and regulations concerning zoning, pollution standards, etc., are being executed daily. For the most part these groups have little access to unbiased scientific consultation. Many air pollution problems require local action and decision. For many pollutants there does not seem to be justification for uniform regulations over large regions. Local distribution and relation to the source are important.

Most people would like to understand the significance and solution to air pollution. This book is written to provide a scientific introduction to air pollution for students who are interested in learning about it as a basis for further research as well as for the laymen and government officials. The significance of the various pollutants is explained in understandable terms and the state-of-knowledge is indicated.

2. Definition of Air Pollution

Contaminants in our atmosphere are called pollutants. These pollutants are in the form of gases, liquid droplets, or particles. An example of a gas is ozone. The term aerosol is frequently used to describe suspended droplets or particles, small enough not to fall rapidly under the force of gravity such as rain drops. Liquid droplets are sometimes classified separately from aerosols, particularly in the case of clouds. Dust is a good example of particles. Generally, the term 'particles' refers to solids larger than aerosols.

Air pollution is not an exact science – not a scientific discipline. It is a multi-disciplinary problem area involving many physical and biological processes. There is no rigorous definition of air pollution; however, it can be thought of as "an undesirable modification of the atmospheric constituents that may harmfully affect flora, fauna, or materials." Thus, to a certain extent whether or not something is an air pollutant is relative to the specific consideration and is arbitrary. Ozone in the lower atmosphere is a pollutant but the natural ozone layer at high altitudes protects the Earth's surface from much of the solar ultraviolet radiation. Whether or not something may harm something else must always be interpreted. The acceptable interpretations will evolve with time as influenced by societies demands and by scientific understanding.

There are both man-made and natural sources of pollutants in the air. Often one of these types of sources completely dominates the other. For some pollutants, natural sources may be most significant in the world wide inventory, while in certain localities, man-made sources predominate.

Detectability of a substance should not be confused with pollution. Many trace elements are absolutely essential to certain flora or fauna. In excess they are pollutants. In general, with a little effort, one can detect most substances at concentrations far below their potential harmful level.

In general, man's senses are not adequate indicators of pollution. His senses are of no avail for many pollutants. Emissions of small particles are generally more of a hazard than large particles, yet large particles are more easily observed by the eye.

Two things have happened to change man's outlook on pollution. The first is that air pollutants have become more sophisticated. It is now recognized that there are many subtle pollutants which adversely affect human beings but man's senses are incapable of detecting or of clearly identifying the source. There is much fear about the potential effects of air pollutants. The longevity has increased to the point where pollutants may have enough time to affect his health. Living and working conditions have steadily improved to where far less quantities of dust and other pollutants are tolerable. The tolerance level for atmospheric pollutants affecting man's comfort, health, and economics has greatly reduced. It is really remarkable that mankind tolerated air pollution for as long as he has. Man's physical ability to tolerate air pollutants probably has not reduced but psychologically he now refuses to accept air pollution.

The original term SMOG probably came from London as early as 1905 to describe the mixture of industrial smoke and natural fog there. It had counterparts in bygone years in Pittsburgh, St. Louis, and doubtless other industrial U.S. cities. The concentration of pollutants could build up to extremely high levels in this type of SMOG, causing illness and deaths. Regulations against incineration, open coal stoves, and industrial coal smoke have almost eliminated this kind of smog in London. The old London 'pea-soup fogs' are distinctly rarer today than in the past.

Today, the term smog usually means to readers and users, 'Los Angeles smog', although it exists in other areas, notably in the western U.S. It has no connection with the old London smog in that neither natural fog nor coal smoke are involved. The Los Angeles smog results from photochemical processes involving sunlight and air pollutants. It is surprisingly difficult to find a formal definition of the Los Angeles type smog. Most writers drift off at the outset into a discussion of the intricate mechanism of its production. According to Berry (1968), "*SMOG is a dispersion of very small droplets, suspended in air, and often containing irritating and corrosive materials.*" In the Los Angeles type smog various nitrogen compounds are found, leading to the characteristic brown color and eye irritation.

The conditions that would favor forming of 'London smog' would destroy the mechanism of production of 'Los Angeles smog' and vice versa. In the London smog there is heavy natural fog and no sunshine while the Los Angeles smog requires sunshine for the photochemical reactions.

Certain key chemicals are classified as pollutants sometimes because of their direct toxicity and sometimes because of their role in forming other more offensive products. Near to the top of any list is sulfur dioxide, a substance that is toxic in its own right and in addition contributes to smog formation.

3. History of Air Pollution

There is a tendency to associate air pollution with the industrial revolution and regard it as a recent problem. Air pollution is very much older than man. Dust has been described as the ubiquitous pollutant of space, and seriously limits astronomical observations. However, closer to home, there are a number of natural forms of air pollution that existed before man and are still very significant contributors to the condition of the air environment. These are allergens, dust, volcanic gases and particles, and smoke. The allergens, most notably rag weed pollen, grass pollen, and molds, are widespread. Their selectivity in attacking only isolated individuals may explain the failure to attempt any control at all of such noxious plants. Man has either suffered with these natural pollutants or changed locales to minimize the unpleasantness.

We are indebted to Dr Munn (author of Chapter 3) for calling our attention to an account of pollution in Detroit in 1762 (Stirling, 1763).

An Account of a Remarkable Darkness
At Detroit, in America: In a Letter From
The Rev. Mr. James Stirling, to Mr. John Duncan:
Communicated by Samuel Mead. Esq; F.R.S.

Detroit, 25th Oct. 1762.
Sir:

A Man in business seldon troubles himself about news; yet the following is so uncommon, I cannot neglect acquainting you therewith. Tuesday last, being the 19th instant, we had almost total darkness for the most of the day. I got up at day break: about 10 minutes after I observed it got no lighter than before; the same darkness continued untill 9 O'clock, when it cleared up a little. We then, for the space of about a quarter of an hour, saw the body of the Sun, which appeared as red as blood, and more than three times as large as usual. The air all this time, which was very dense, was of a dirty yellowish green colour. I was obliged to light candles to see to dine, at one O'clock, notwithstanding the table was placed close by two large windows. About 3 the darkness became more horrible, which augmented untill half past 3, when the wind breezed up from the S.W. and brought on some drops of rain or rather sulphur, and dirt, for it appeared more like the latter than the former, both in smell and quality. I took a leaf of clean paper, and held it out in the rain, which rendered it black whenever the drops fell upon it; but, when held near the fire, turned to a yellow colour, and when burned, it fizzed on the paper like wet powder. During this shower, the air was almost suffocating with a strong sulphurous smell; it cleared up a little after the rain. There were various conjectures about the cause of this natural incident. The Indians, and vulgar among the French, said, that the English, which lately arrived from Niagara in the vessel, had brought the plague with them. Others imagined it might have been occasioned by the burning of the woods: But I think it most probable, that it might have been occasioned by the eruption of some volcano, or subterraneous fire, whereby the sulphurous matter may have been emitted in the air, and contained therein, until, meeting with some watery clouds, it has fallen down together with the rain.

I am, Sir,

Your most humble servant,
James Stirling.

This section makes no attempt to provide a technical history of air pollution practices, laws, abatement, and development of the state-of-the-art. Such a discourse would fill a lengthy book on its own merits. Also, it is not obvious that such a historical review would readily aid in making any future decisions. In the appropriate chapters, temporal changes of specific pollutants and effects are discussed where ever they are known. Rather, this section provides a philosophical discussion of some historical events, which indicates how society has changed its demands and overview of pollution.

As society progressed and man was able to contribute to the air pollution, the main pollutants were probably smoke from camp fires, dust from cultivation, and noxious odors. Before the industrial revolution, these were short range irritants and for the most part the source was clearly obvious. The solution for discomfort was to move your position. Man's long toleration of smoke, dust and odors is remarkable. Mere bad odors appear to have been tolerated for a long time, outright toxicity being required for condemnation. The correlation of noxious odors and potential harm to man is not easy to make. In some communities you can still predict the weather and determine the wind direction by the odors.

Of the older and perhaps lesser pollutants, one of the authors (R.V.) offers the following childhood recollection as of 1915 from San Francisco (a remarkably clean-air city, thanks to the prevailing westerly winds). When the wind blew from the north-west, a miserable odor of burning cocoa husks from the Ghirardelli chocolate factory (now a famous tourist and shoppers' mecca), suffused the city. A northeast wind at certain seasons blew the surprisingly unpleasant odors of the North Beach canneries over the city, probably an odor of stewing tomatoes. Regularly on September 15, when the prevailing westerlies subsided for a brief interval at the end of the summer, a genuine smog covered the city with a marked and acrid odor of coffee roasting accompanied with a burning rubber tang. At still other times, the stock yard odors blew over the city, but in San Francisco, this odor was negligible in magnitude compared with its pungency in Chicago, St. Louis, and Kansas City.

In seaside areas, salt spray blows into the atmosphere fairly continuously, obviously augmented during storms. The salt condenses on electric insulators and frequently causes breakdown and power failure during the first rainfall after a long dry spell. The problem extends inland; thus the 270 thousand volt transmission line from Hoover Dam to Los Angeles is constructed in duplicate so that the insulators on one or the other set of lines can be continually subject to cleaning. Salt spray and/or fog have been suspected of causing sinus inflammations. They are, in some places and on some occasions, accompanied by plankton sprayed up with the salt, and these are highly irritating. The subject of pollutants, both natural and man made, must include reference to radioactive fallout. The existence of an emotional and even hysterical reaction to fallout has to some degree obscured some significant underlying facts. The first is that radioactive fallout from natural causes occurred long before nuclear explosions. Thus if a person wiped off a radio antenna with a clean cloth following a strong wind storm, and then brought the cloth near a nuclear radiation detector – for example, in bygone years, a charged electroscope – the cloth showed appreciable radioactivity. A good natural background level must be known against which to measure man-made radiation levels.

Over the last 100 yr or so some of the more toxic pollutants have been controlled by laws. Few people realize that the expression "mad as a hatter," or Alice in Wonderland's friend, the mad hatter, arose from the use of mercury in the manufacture of felt hats. Mercury makes the felt fibers stand upright. The long-time toxic effect of inhaling mercury vapor which evaporates from the felt is irrational behavior. Hatters notoriously went mad. This usage of mercury is generally prohibited by law.

An early industrial chemical process in England was the manufacture of sodium sulphate which was then roasted with limestone, ultimately resulting in sodium carbonate. A stack effluent from this process was hydrochloric acid gas, which in due course dissolved in the dew on meadow grass; the cows that ate the grass got sore mouths from the hydrochloric acid. By 1864 the discharge of hydrochloric acid into the air was banned by law, thereby ruining – it was said – the thriving industry for the production of sodium sulphate. Today, the trapped hydrochloric acid is the primary product and sodium sulphate is a secondary by-product. In early years, following the prohibition on discharging HCl, the trapped and dissolved hydrochloric acid was then taken to sea and dumped.

There are many examples wherein with some time and effort and redesigned processes, pollutants can be utilized into useful products and overall efficiency improved.

4. Complexities of Air Pollution

Air pollution is very complex. Its investigation requires the efforts of many disciplines and many other multidisciplinary areas. Air pollution is also tied to water and soil pollution. Some sources of atmospheric pollution come from bacteria in water or soil and from evaporation. On the other hand, water and soil may be a sink for atmospheric pollutants. However, detailed consideration of water and soil pollution is beyond the scope of this book. The disciplines include biology, chemistry, geology, and physics. Some of the broader research areas include agronomy, ecology, geophysics, limnology, medicine, meteorology, and oceanography. Focus on a single discipline or research area is not likely to play a key role in reducing air pollution.

The scientific study of atmospheric pollution is covered in four main parts in this book. The first describes the types and sources of air pollutants. Man-made and natural sources are described. The inventory of pollutants is given where available. The sinks of pollutants are discussed. Data on monitoring the various pollutants are very sparse. It is very difficult to compare current observations with measurements taken only 5 to 10 yr ago because of changes in techniques. It is especially difficult to try to estimate the long term trends, i.e., has a pollutant increased over the last 20, 50, or 100 yr.

The natural environment is described and the perturbations from the SST, gasoline vehicles, volcanoes, and industry are estimated. The interactions of the various pollutants in the atmosphere – the atmospheric chemistry – are provided.

The second part of the book describes the transport of pollutants, their flow and dispersion. Meteorological effects on pollutants and vice versa are described. The distribution of air pollutants is important as there is no such thing as a uniform distribution.

The third part of the book concerns the effects of air pollutants. The effects on human beings and vegetation are described in detail.

Fourth, air quality criteria, monitoring techniques, and abatement techniques are described.

Just as there are uncertainties in predicting elections, horse races, and sporting events, there are uncertainties in predicting the exact behavior and life cycle of most pollutants and in predicting their effects. Just as physicians will disagree on the diagnosis and treatment of a patient, the air pollution 'experts' will disagree on the allowable limits of a specific pollutant, on its life cycle, effects, etc. We do not even know the natural composition of the atmosphere well enough, especially the minor constituents. The chemical reactions in the natural atmosphere are very complex and many are undetermined. Air pollution is a multidisciplinary problem area, not a discipline which may be described by some exact rules, principles, etc. Air pollution will never be exactly defined; it is relative to society's demands which will shift with time. This does not mean that logic and scientific techniques and principles cannot or should not be applied to air pollution. They should be and can be applied even much better in the future than in the past. However, it is important to remember that there are no exact unchallengeable answers for the description of air pollutants or for their elimination.

Predictions of overall future implications of the injection of various pollutants are particularly risky. However, such predictions must be made and calculated risks must be taken. A goal of no man-made air pollutants is unrealistic and not achievable. Catastrophic effects of air pollutants are periodically predicted. Questions are often raised about triggering a mechanism – a non-linear effect, wherein a minor perturbation may lead to a catastrophy. It is too complicated to make predictions many years into the future. One must often be content with making some analogy with natural phenomena. In other cases one can carefully measure the effects and accumulation of specific pollutants and if they get out of bounds, make changes. This is what is really happening in the case of sulfur coal and fuel oil in populated areas. The accumulation of sulphur compounds in the atmosphere was unacceptable, resulting in standards on the amount of sulphur in such fuels. Earth's air environment does not appear to be precipitously balanced ready to be ruined by the push of a feather. A goal of reducing the current injection of pollutants and minimizing the overall harm is achievable.

References

Berry, R. S.: 1968, *Encyclopedia Britannica.*
Stirling, J.: 1763, *Phil. Trans. Royal Soc. (London)* **53**, 63–64.

ATMOSPHERIC POLLUTANTS

ROBERT VARNEY and BILLY M. McCORMAC

Lockheed Palo Alto Research Laboratory, Palo Alto, Calif., U.S.A.

1. Introduction

In this chapter, the basic characteristics of atmospheric pollutants will be presented including their character, methods of dispersion or accumulation, sources, sinks, and reactions. Some comments concerning their characteristics that warrant their classi-fication as pollutants will be offered.

There are many claims of an air pollution doomsday; however, there do not appear to be any serious arguments that we are approaching a catastrophic failure of our atmosphere to support human beings. Processes in our atmosphere are poorly under-stood, however. Many aspects of atmospheric pollution cannot be properly evaluated without giving consideration to the oceans and solid earth as well as to the total energy input from the Sun and the influence of the total biosphere. Over billions of years, Earth has not melted or frozen, during which time the biosphere has evolved in complexity. During this period of time, there have been many major perturbations, such as volcanic eruptions.

There is much evidence to indicate that the worst has been reached on air pollution. There are also arguments that many urban areas have seen their worst incidents, additional emission would only increase the number of bad days. In many urban areas, improvements in air quality have been documented for 10 to 20 yr. In others, much improvement has occurred in the last few years. Except in newly developing areas, for the most part atmospheric pollution is decreasing. This does not mean that we can relax.

Most man-made pollutants are a local problem. Deposition depends on variations in the source and meteorology. The serious effects are limited to about 50 km distance or much less. A great number of polluted local regions could each contribute a small amount to more distant regions thereby creating significant pollution levels at distances much larger than 50 km. However, if the local problems are solved – if standards are met for the persons living within a few km of the local source, then there will not be a long distance problem. The fact that Earth is not facing an air pollution doomsday is of no comfort to the person suffering from local pollution.

There are very few long distance or world-wide atmospheric pollutants. The most obvious examples are CO_2 from burning fossil fuels and particulate matter from very large volcanic eruptions.

Most industrial hazards, such as Be dust, toxic gases, etc., are subject to safety regulations and inspections by one or more governmental safety organizations. As such hazards are recognized (Sax, 1963), adequate control is straightforward and there is little mystery and no excuse for failure. More attention in the future should be given to asbestos and heavy metal hazards. New materials and processing techniques may produce new hazards.

At the outset, it is important to recognize that some of the worst and most widespread pollutants are not man-made nor are they at all readily controllable by mankind. It has been said that the explosion of Krakatau in 1883 put more particulate matter into the atmosphere than mankind's smoky fires throughout all times (Abelson, 1971). Volcanic fumes containing SO_2 and H_2S pour from the ground continuously, augmented on occasions by greatly increased emissions during violent eruptions. Forest fires, ignited by lightning, must be considered to be natural sources of pollution. Decaying vegetation that evolves CH_4 and hydrocarbon fumes from open tar pits are other natural pollutants. Clouds and fog are so commonplace that they are rarely classified as pollutants, although a pilot attempting a landing in fog surely regards the fog as just that. Finally, the allergens, pollens of weeds and wild grasses, molds, even ocean spray containing plankton, cause misery to vast numbers of people.

Man-made pollutants are of more immediate concern because of the possibility and hope that they can be controlled and even eliminated. We are anxious to point out at once that atmospheric pollution by man is literally an ancient problem. Smoky fires have burned for untold centuries. The earliest history of the discovery, mining, and purification of metals dates from about 4000 BC. Then, and today, the arts of metal production and fabrication have caused the generation of noxious and toxic fumes. With the rise of industry, these man-made fumes became more important, but technology and laws have reduced their significance. However, man-made atmospheric pollution has become important for two reasons. The first results from population growth; the growth in demand for manufactured products and the growth in demand for electric power have multiplied even low levels of pollutant output to major problems. Thus for example, coal burning in a large urban area for space heating and electric generation, even with modern restrictions on the quality of allowable coal and the use of automatic burning controls, leads to an aggregate of SO_2, fly ash, and other contaminants that is severe yet cannot be blamed on any single source. The second modern atmospheric pollution problem is clearly associated with the enormously increased use of motor vehicles. Here again, we are concerned with outputs from single vehicles measured in parts per million (ppm), yet with millions of cars in operation, the aggregate is critical in certain locales. Technologists of a century ago would have been aghast at the suggestion that contamination measured in ppm might ever be a concern and at the thought of attempting to regulate such low levels of output, or in fact even measure them.

With the growth of urban centers, industry, and technology, new aspects of atmospheric pollution are continually being created. A century from now, this chapter will doubtless be as amusing and as out of date as the problem of HCl fumes cited in Chapter 1 must seem today. With the growth of both population and potential pollutants, greater control of output than simple remote location of factories and power plants will be necessary.

Air contaminants run the gamut of solids, liquids, and gases. The historic London smog described in Chapter 1 is a mixture of solids and liquids. Sulfur dioxide, one of the worst pollutants, is a gas but reacts to form H_2SO_4 which is a liquid; this is, if anything, even worse. We offer a brief classification of the various forms encountered.

Solids, often spoken of as particulate matter, are probably most commonly either forms of soot, i.e., carbon with contamination, or dust, including volcanic dust, i.e., largely SiO_2. Ice must not be overlooked among the particulates. Allergens are of course far more complex solids. The size, shape, and texture of particulates are extremely variable. Despite the predilection of theorists for ideal, spherical particles, sharp, even needle-like particles are known. Volcanic solid dust was noted by Humphreys (1940) to consist mainly of thin-shelled bubbles or fragments thereof. Fly ash may also involve thin-shelled bubbles. These bubbles are sometimes called 'cenospheres', the prefix, a Greek derivation, denotes that the spheres are empty.

Liquids may preserve a much more nearly spherical form, at the worst spheroidal. Like solids, there is almost no limit to their size ranging up into mm in the case of raindrops and fading off into mists below 0.05 μm diam.

Certain terms recur in connection with both solid and liquid matter. Two common ones are 'condensation nuclei' and 'aerosols'. The term 'condensation nuclei' arises from the fact that the growth of drops often goes in stages, from a few molecules to a small but observable droplet, and from a droplet to an actual raindrop size. In both cases, some unusual starting nucleus, unusual compared with the simple molecules O_2, N_2, or even H_2O, is found to initiate the growth. Thus even a molecule of salt sprayed from an ocean breaker may start the growth of a future fog mist particle. Sulfuric acid is notably hygroscopic, and 'sulfuric acid mist' is familiar in chemical laboratories and plants. In some cases, a single ion of any molecule may serve as a starting nucleus. The details are unbounded. Growth of larger drops doubtless starts from intermediate particles or droplets although the growth of raindrops is far from being a completely resolved or understood subject. Hailstones clearly display a shell-like structure disclosing a history of growth. In most cases, the idea of a seed nucleus or condensation nucleus is a useful one.

Aerosols, as used in atmospheric studies, are particles or droplets that float in the atmosphere. Since solids or liquids with specific gravity as low as that of air are unlikely, including even the volcanic bubble particles, the term 'float' needs further explanation. The matter may 'float' because of updrafts of air. Such currents can lift large hailstones under some conditions. In still air, floating may simple be an illusion arising from a very slow rate of fall. Lamb (1970) presents a table of computed rates of fall of particles, assumed spherical in shape, of various sizes, at various heights, following a modified Stokes formula. Thus at 10 km altitude, a sphere of density 2.3 g cm^{-3} (a good value for silica dust) with a diameter of 5 μm would fall 1 km in 12.5 days whereas a particle ten times smaller, or 0.5 μm in diameter would take 62 weeks for the same descent. Lamb does not carry the calculations below 10 km claiming that matter is washed out by rain long before it has time to fall very far. He also cuts off calculations at particle sizes of 0.5 μm on the grounds that below this size, light scattering from the particles fades off into that of the ordinary atmosphere.

It is clear from Lamb's analysis that ordinary clouds appear to float since their rate of fall is too slow for un-instrumented detection. Occasionally, authors attempt to define an aerosol as any particle under 1 μm in diameter. While the choice is understandable, the combination of various factors like shape and density of the particles makes such a definition unduly arbitrary. The concept of suspended or nearly suspended matter in the atmosphere is, however, a useful one for certain considerations.

Smog has been defined in Chapter 1 with primary emphasis on its role in reduction of visibility; its secondary identification characteristic is irritation, particularly of mucous tissue. In the ensuing analysis, reference will be made from time to time to primary and to secondary pollutants. The distinction is reasonably obvious and is best brought out by an example. Sulfur dioxide is a primary combustion product and effluent from coal-burning furnaces. It would thus be a primary pollutant. It can react in time to form H_2SO_4, first from oxidation even by the oxygen in the air to SO_3 followed by absorption of water from the atmosphere to form the end product of H_2SO_4. The sulfuric acid is a pollutant in its own right but is called secondary because of its dependence on the primary formation of SO_2.

It was indicated at the start that the grounds for classifying an additive in the atmosphere as a pollutant needed to be declared. Clearly, the most critical consideration is whether it is actually toxic to human beings. In the analysis of chemicals later in this chapter, an attempt will be made to classify each chemical in this regard. It is of interest to note a variety of aspects of toxicity. Thus there are chemicals that are toxic in extremely low concentrations, whereas others, like HCN for example, can kill above certain levels but are not particularly harmful at lower concentrations. There is an account from World War I that the French were intent on developing HCN as a war gas until a scientist, to prove his point, confined himself and a dog in a sealed room, and turned on a tank of HCN until the dog dropped dead – and then emerged in good health himself. There are chemicals whose toxic action is rapidly dissipated by elimination from the body, and there are others that are augmented by accumulation over long periods. Cumulative poisons are of course dangerous at low levels of concentration if the level persists for a long time.

Next to actual toxicity, irritation to the skin, eyes, and respiratory tract is perhaps the most important qualification for naming a substance as a pollutant. There is an overlap between toxicity and irritation in case the irritation reaches damaging levels. Since the primary, present-day concern is not with actually killing levels, irritation is the first severe debilitating phenomenon of pollutants.

Destruction of both vegetation and inanimate material is often a feature of pollutants. Damage to cars, buildings, paint, and other objects, simply by corrosion, is observed in growing frequency. There is the economic loss due to higher cleaning costs and shorter life.

Radioactivity as a feature of pollutants requires brief discussion as it is a hazard, on the one hand, and, in addition, does exist in natural form.

The reduction in visibility is a significant qualification of an air additive as a pollutant, obviously ranging over great extremes of severity. The average person estimates air pollution on the basis of visibility. Low-level smogs (excluding their irritating characteristics) may be only a nuisance, whereas pea-soup fogs are major hazards.

At the lowest level, pollutants may be nothing more than annoying. While minor annoyance is scarcely grounds for great concern, it becomes serious as the number of people who are annoyed increases. Finally, there is the very common concern that today's annoyance grows into tomorrow's more severe damage and might be more easily controlled while it is still only at the lower level.

In analyzing pollutants, one tool of study is their origin or mechanism of production. The current point of view is that once a severe smog has formed, covering several hundred square miles, its dissipation by human actions is unthinkable. The concern is therefore to determine the origin and cause, and stop the smog before it is formed rather than dissipate it after its formation.

The removal of pollution from the atmosphere is first and foremost done for us by rain, wind, and simply volume dilution. Some of the pollutants fall into the ocean and are absorbed. Insofar as they fall on land, as in cities, they may constitute a new nuisance. Chemical processes in the atmosphere are of interest but they often produce new, secondary pollutants rather than 'clear' the air.

Volume dilution is covered in detail in Chapter 3. In general, dangerous concentrations of effluent result when a ceiling exists through which the stack effluents will not penetrate. Such a ceiling is found from time to time. Air temperatures normally fall with increasing altitude, but under certain conditions, the air temperature rises from the ground to some certain height and then declines with further increase in altitude. The altitude at which the temperature change reverses is called 'the inversion height' and the transition zone is called 'the inversion layer'. Gases such as effluents can rise by convection to the inversion layer but are than trapped and can neither rise nor fall. Typical inversion heights are found from 100 to 1000 m. A low-lying inversion layer traps effluent gases or smokes into a greatly reduced volume. In addition a near-zero wind speed is required to obtain high concentrations of pollutants.

According to Lamb (1970), particulate matter such as volcanic dust persists in the atmosphere for a long period, like several years, only if it is initially projected upward through the tropopause. The tropopause is defined as the altitude at which the decline in air temperature with altitude levels off. It is the lowest of several such reversals, and it occurs at about 10 km in arctic areas and at about 20 km in tropical areas. A volcanic eruption in the tropics must therefore be more violent than an arctic one to project dust into trapping altitudes. Above these troposphere altitudes, the dust migrates around in the prevailing winds, crossing the equator without much sign of an obstacle. It is subject to up and down drafts and to the slow fall predicted by Stokes's law but not to the scrubbing action of rainfall at these altitudes. Dilution is highly effective, but the mass of particulate matter from some volcanic eruptions has been observable and colored the sunsets all over the earth for periods ranging up to three years.

While some uncertainty exists, it is likely that once the particulate matter falls below the tropopause it can serve as condensation nuclei for raindrops. What is known is that the particulate matter vanishes rapidly coincidentally with rain and that raindrops condense and form more readily with the aid of condensation nuclei.

Chemical processes will be treated in detail under the heading of each common chemical pollutant separately, but at this point, the biological removal of CO_2 is described briefly to illustrate an aspect of chemical removal problems still remaining to be solved. Basically, plants are known to absorb CO_2 and to evolve oxygen. There is preliminary evidence (Bolin, 1970) that as the concentration of CO_2 rises, the efficacy of its removal by plants actually increases. Whether it does or not is of great importance to know in the attempt to understand the overall equilibrium of atmospheric pollution, for if processes exist that rise to meet increased pollution loads the task of control is greatly aided. Before tackling the problem of equilibrium, we must first consider the division of the atmosphere into regions.

2. Natural Atmosphere

Pure air does not exist. Instead there is a dynamic exchange between the atmosphere, solid earth, hydrosphere, biosphere, etc. Approximately 78 % by volume of the clean dry atmosphere consists of the relatively inert gas N_2. Next O_2 constitutes about 21 % and all other constituents account for 1 %. The total mass of the atmosphere is 5.3×10^{21} g. Thus, there is about 10^3 g of air above each cm^2 of the Earth's surface; 90 % is below 16 km altitude, 99 % below 30 km, and 99.9 % below 48 km. A list of the

TABLE I

Constituents in clean dry air

Constituent	% by volume	% by weight
N_2	78.088	75.527
O_2	20.949	23.143
Ar	0.93	1.282
CO_2	0.0318	0.0456
Ne	1.8×10^{-3}	1.25×10^{-3}
He	5.24×10^{-4}	7.24×10^{-5}
CH_4	1.4×10^{-4}	7.75×10^{-5}
Kr	1.14×10^{-4}	3.30×10^{-4}
N_2O	5×10^{-5}	7.6×10^{-5}
Xe	8.6×10^{-6}	3.90×10^{-5}
H	5×10^{-5}	3.48×10^{-6}
NO_2	1×10^{-7}	3×10^{-7}
O_3	2×10^{-6}	6×10^{-6}
SO_2	2×10^{-8}	9×10^{-8}
CO	1×10^{-5}	2×10^{-5}
NH_3	1×10^{-6}	1×10^{-6}

atmospheric constituents for clean dry air is found in Table I (*Handbook of Geophysics and Space Environment*, 1965). The density of air as a function of altitude is shown in Figure 1.

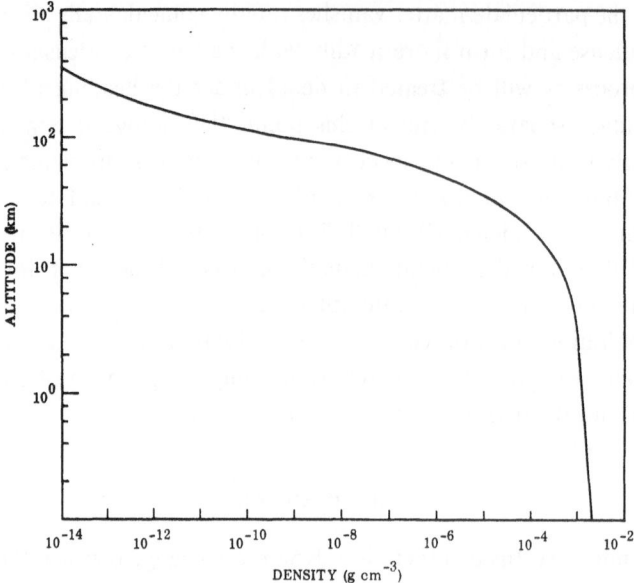

Fig. 1. The average density of air as a function of altitude (Handbook of Geophysics, and Space Environment, 1965).

Observations of O_2 over 50 yr fail to show any detectable change in its concentration (Machta and Hughes, 1970). It is believed that essentially all, if not all, of the O_2 in the atmosphere is of biological origin. The basic process is photosynthesis:

$$CO_2 + H_2O + h\nu \rightarrow CH_2O + O_2. \tag{1}$$

The oxygen has been identified as coming from the water in this process. The O_2 is removed from the atmosphere by decomposition, the reverse process of Equation (1), conversion to O_3, oxidation of FeO, and oxidation of sulfides. An excellent description of the detailed oxygen cycle is given by Cloud and Gibor (1970). The quantity of N_2 also seems to be constant. Most of the minor constituents vary considerably from time to time and place to place. There are seasonal and diurnal variations. The regions are different from the lower latitudes. Carbon dioxide is observed to be gradually increasing. There is a serious lack of understanding about the role of many of the minor species in atmospheric processes. The gases Ar, Ne, Kr, and Xe can be ignored in air pollution studies as they seem to present no hazard and play no role in air pollution.

Discussions of atmospheric pollution involve consideration of concentrations measured in ppm and solar photochemical processes that require atmospheric thicknesses measured in km. In pursuing these studies, there is a tendency to forget that our atmosphere is subjected to much more drastic effects. The change from day to night alone is of very large magnitude. The change in mid-latitudes from winter to summer is enormous. The climatic difference in the atmosphere between that over the shores of Hudson Bay and Scandinavia, both at about the same latitude, is very different. The coastal region atmosphere bears little resemblance to that of desert country.

It is useful to describe the various regions of the atmosphere and their characteristics. There are a number of regions, as shown in Figure 2, which reflect the kinetic temperature profile in the atmosphere (*Handbook of Geophysics and Space Environments*, 1965).

The troposphere is the region just above the Earth's surface which has an almost uniform decrease of temperature with altitude amounting closely to 6.5 km^{-1}.

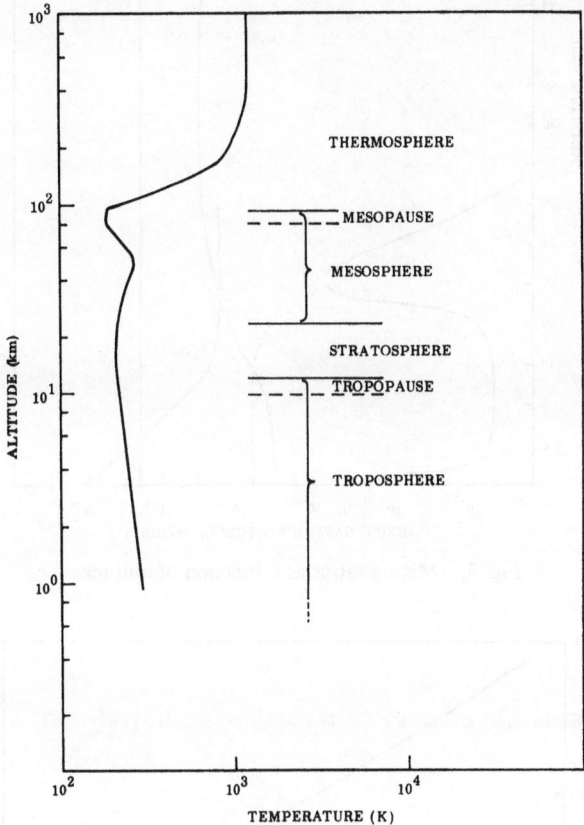

Fig. 2. Atmospheric temperature as a function of altitude. The temperature zones are named.

Temperature inversions are often found in the troposphere as described in Chapter 3. The tropopause is the boundary at the top of the troposphere where the temperature remains constant with altitude. The weather region is in the troposphere with high winds and the highest cirrus clouds being found in the tropopause. The altitude of the tropopause varies much with time but is highest, about 20 km, at the equator and lowest, about 10 km, at the poles. There is a break in the tropopause in the mid-latitudes.

The next higher region is the stratosphere which is approximately at a constant temperature. The stratosphere is thicker over the poles than at the equator. Above the stratosphere the temperature increases to a maximum at the middle of the meso-

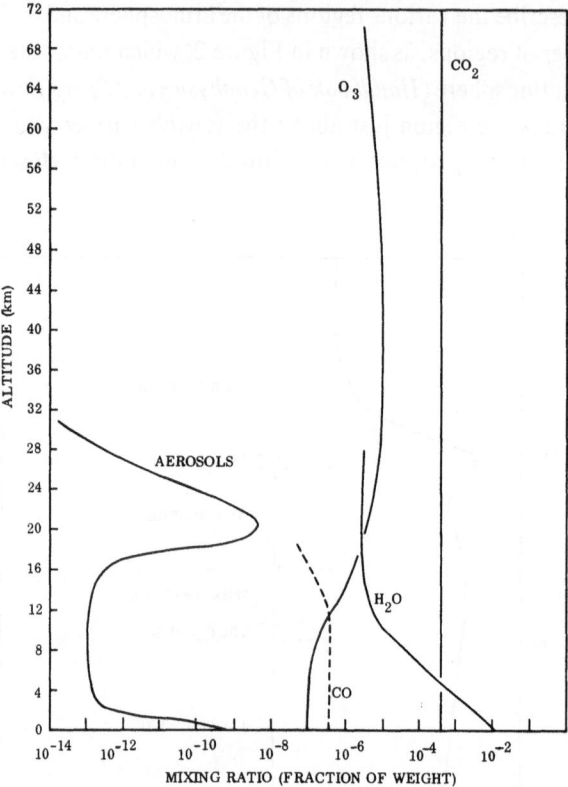

Fig. 3.   Mixing ratios as a function of altitude.

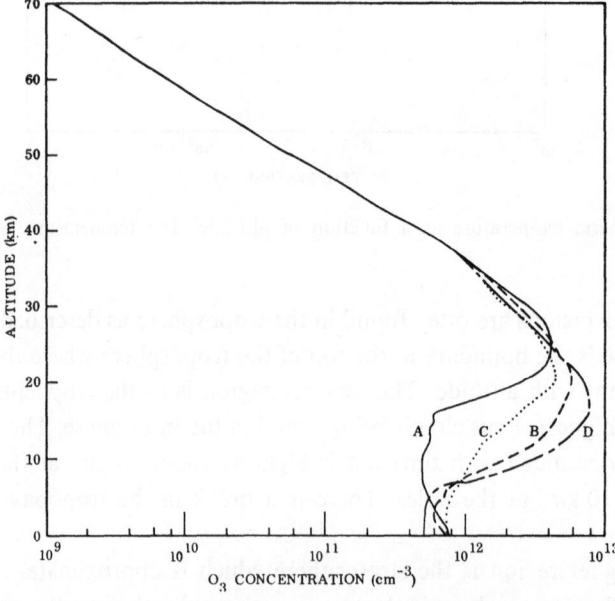

Fig. 4.   Concentration of O_3: A – Tropical, B – Mid-latitude winter, C – mid-latitude summer, D – subarctic winter (McClatchey et al., 1970).

sphere and then drops back to a minimum value at the mesopause. Finally above the mesopause is the thermosphere where the temperature increases with altitude.*

The atmosphere also contains aerosols and H_2O. These are especially highly variable with time and location. The average composition of the troposphere, stratosphere, and lower mesosphere for aerosols, H_2O, CO_2, O_3, and CO is shown in Figure 3 as a function of altitude. The mixing ratio, fraction of the total, for CO_2 is independent of altitude while it is variable for the others (*Handbook of Geophysics and Space Environments*, 1965; Newell, 1971). The column density of H_2O above the Earth's surface varies by a factor of 10 from time to time and place to place; the average at the equator is 4 to 5 times higher than at the poles. The column density of O_3 can vary by a factor of 2 at the same location from day to day.

The injection of H_2O by man into the troposphere is small and unimportant; although man can influence the local humidity or climatic conditions by large scale irrigation or with a large artificial lake.

In order to visualize the O_3 layer in the stratosphere it is necessary to plot the concentration of O_3 per unit volume as a function of altitude as shown in Figure 4 (McClatchey *et al.*, 1970). Note the large variation in O_3 at the same altitude depending upon season and latitude.

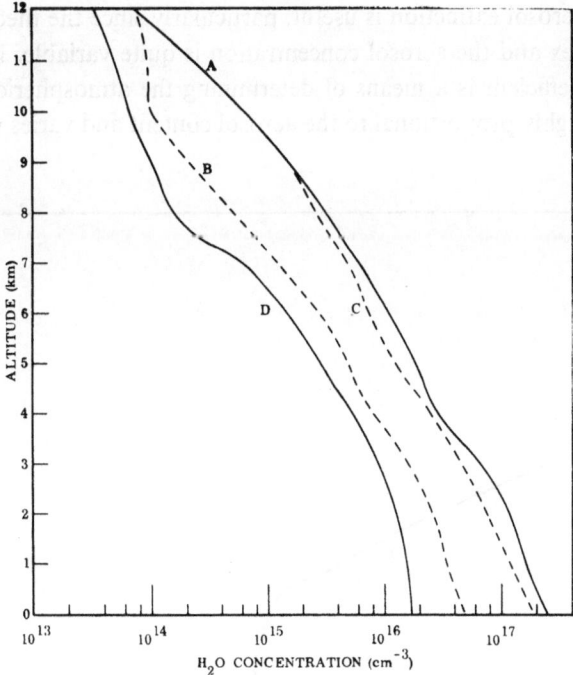

Fig. 5. Concentration of H_2O: *A* – Tropical, *B* – Mid-latitude winter, *C* – mid-latitude summer, *D* – subarctic winter (McClatchey *et al.*, 1970).

* Readers will find other names used for various altitude bands. Reference may be made to encyclopedic books like the *Handbook of Geophysics and Space Environments* (1965).

Figure 5 shows the seasonal and latitudinal variation in H_2O which is quite significant.

A. EXTINCTION COEFFICIENT

One of the most obvious manifestations of atmospheric pollution is reduced visibility. Aerosol particles are particularly important in affecting the visibility. As light is transmitted through the atmosphere it is absorbed and scattered out of the line of sight (extinction) and is scattered into the line of sight (multiple scattering). In the presence of aerosols extinction dominates multiple scattering. The light intensity relationship, ignoring multiple scattering, is

$$I = I_0 \exp\left(-\int_0^x \beta_x \, dx \right), \tag{2}$$

where I is the intensity that would be observed at a distance x from a source I_0 and β_x is the altitude dependent extinction coefficient. The extinction coefficient has four components: absorption and scattering from aerosols and from gases. It has not been possible to date to independently measure all four; however, evidence indicates that in the presence of some aerosol content aerosol scattering dominates. An approximate estimate of the aerosol extinction is useful, particularly since the theory of scattering of light is complex and the aerosol concentration is quite variable. Measurement of the extinction coefficient is a means of determining the atmospheric pollution. The value of β_x is roughly proportional to the aerosol content and varies with wavelength

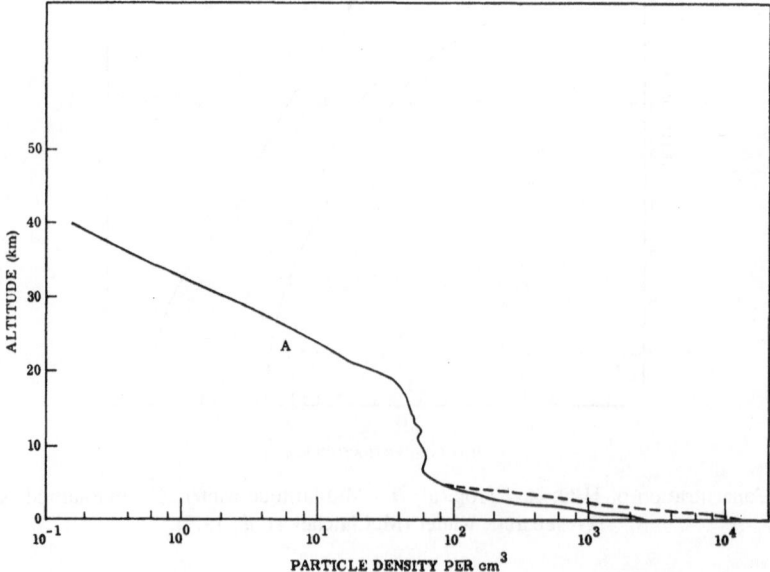

Fig. 6.   Particle density of aerosol: A – clear visivility, B – hazy visibility (McClatchey *et al.*, 1970).

as λ^{-1}. The value of β_x in the stratosphere is the order of 10^{-4} to 10^{-3} km^{-1}. It is much larger near to the surface being about 10^{-1} km^{-1} for a clear atmosphere and about 1 km^{-1} for a hazy atmosphere. After volcanic eruptions the extinction coefficient may increase by a factor of 10 or more.

For a model atmosphere, McClatchey *et al.* (1970) have established two aerosol models. They differ in the altitude range of 0 to 5 km. One is for a clear atmosphere (visibility 23 km) and the other is for a hazy atmosphere (visibility 5 km). The aerosol model assumes that only particles of 0.02 to 10 μm are present. The concentrations for clear and hazy atmospheres are shown in Figure 6. The aerosol number density as a function of particle radius $N(r)$ is given by

$$\frac{N(r)}{\Delta r} = 0.9 \times 10^{-13}\, r^{-4} \quad \text{for} \quad 0.1\ \mu\text{m} < r < 10\ \mu\text{m}$$

$$= 0.9 \times 10^{-9} \qquad \text{for} \quad 0.02\ \mu\text{m} < r < 0.1\ \mu\text{m}.$$

(3)

A handy rule has been found by Charlson (1969) to relate visibility and aerosol content when the humidity is less than 70%, i.e., $LM = 1.8 \times 10^{-2}$ g cm^{-2}, where L is the meteorological range in km and M is the mass concentration in g cm^{-3}.

B. ATMOSPHERIC OXYGEN SUPPLY

There have been a number of contentions that man may destroy his oxygen supply and thus himself through pollution. For example, it is popular to suggest that burning the fossil fuels will deplete the atmospheric oxygen.

Earth's atmosphere is 20.946% O_2. The total moles of O_2 in the atmosphere are obtained by the product of the moles cm^{-2} times the area of the Earth. The moles cm^{-2} are:

$$\frac{1.01 \times 10^6 \times 0.20946}{980.62 \times 32} = 6.7 \text{ moles cm}^{-2}.$$

Since the area of the Earth is 5.1×10^{18} cm^2 there are 3.4×10^{19} moles of O_2 in the atmosphere.

The burning of fossil fuels can be expressed as

$$C + O_2 \rightarrow CO_2.$$

(4)

Additional O_2 is consumed by H. Since saturated hydrocarbons have the chemical form $C_N H_{2N+2}$, and since the end product of burning is H_2O, it results that roughly half as much O_2 is used up by the H as by the C. Thus we see that 1 to 1.5 moles of O_2 are consumed for each mole of C.

The amount of fossil fuels is estimated to be 7.5×10^{16} moles of C (Environmental Quality, 1970). This would consume 7.5×10^{16} moles of O_2, which is 0.23% of the atmosphere; a value of 0.34% results from adding O_2 for the H. There are certainly pollution problems associated with the burning of fossil fuels but one of them is not depletion of Earth's oxygen. The amount of H_2O in the stratosphere is about 3×10^{12} kg. By estimating a residence time of 2 yr for H_2O in the stratosphere, one can calculate that 5×10^4 kg s^{-1} of H_2O are entering the stratosphere from high altitude thunderheads.

3. Equilibrium of a Reactant Substance

One of the problems about atmospheric constituents, especially contaminants, is the mechanism by which each constituent is formed and by which it is destroyed. A clue to this sort of information can sometimes be obtained by studies of the changes in concentrations of the substance with time and with real or artificial changes in rates of production. The changes as a function of time almost invariably satisfy a linear, first order differential equation. It is often possible to derive critically needed information from the equation without even solving it. The constants in the equation then point to the mechanisms that may be involved. This procedure is illustrated with some examples that involve progressively more complicated conditions.

Suppose first that a substance in the atmosphere is being introduced, by natural or artificial processes, at a constant rate α and that it is not being destroyed at all. The constant α may be in molecules s^{-1} or tons yr^{-1} or any other useful units. The differential equation involved is then nearly trivial; it reads

$$\frac{dx}{dt} = \alpha \qquad (5)$$

where x stands for the concentration of the substance in the atmosphere at time t. The solution of this equation is

$$x = \alpha t + C \qquad (6)$$

where C is a constant of integration and represents the level of the substance in the atmosphere at time $t=0$.

If the same constituent or contaminant is also destroyed at some constant rate β then the differential equation becomes

$$\frac{dx}{dt} = \alpha - \beta \qquad (7)$$

and the solution is

$$x = (\alpha - \beta)t + C. \qquad (8)$$

If the assumption that α and β are really constants is correct, and if it is found that the concentration x of the substance in the atmosphere is not changing with time, the value of dx/dt is zero, and $\alpha = \beta$. The atmosphere is in a steady state, and the rates of production and of destruction of the substance under study are equal.

Suppose next that man starts to produce more of the substance and introduces it into the atmosphere. He thus increases the value of α, but for the moment, we picture

α still to be constant but at a new and higher value. Then dx/dt is no longer zero, but the solution of the differential equation is still given by Equation (8). With α greater than β it is seen that the concentration x rises with time. If in fact the concentration of a pollutant is noted to be rising uniformly and continuously, one can conclude that the formulation described above is possibly correct, although only the difference $\alpha - \beta$ can be determined.

The sort of linear rise with time just described may be slow and almost undetectable, but since it is cumulative, the concentration will continue to rise inexorably with time.

Fortunately, if the destruction process measured by the coefficient β is a typical chemical process, β is not constant but grows larger in proportion to x. In other words, the more of the substance there is in the atmosphere, the faster the reaction processes will operate to remove it. Mathematically, this situation would be described by writing

$$\beta = b\,x,\text{*} \tag{9}$$

indicating that as x gets larger, β also gets larger. The coefficient b is a new one and is constant according to the present assumptions. The differential equation then takes the form

$$\frac{dx}{dt} = \alpha - b\,x \tag{10}$$

and its solution is

$$x = \frac{\alpha}{b}e^{-bt} + C. \tag{11}$$

This equation pictures a very different state of affairs. First, it discloses that if man and nature do not change the rate of production α of the constituent for a long time, the concentration x ultimately will settle to the constant value C. The value of C can be read from the differential equation by inserting that $dx/dt=0$, i.e., that x has become constant, and the constant value of x is just α/b. Should mankind now raise the value of α to a new but constant value, the pollutant would not simply go on growing in concentration forever but would level off after a time at α/b as before but with the new value of α. The only hazard here would arise if man not only generated more pollutants in the atmosphere but each successive year generated more than the preceding year. This equation shows that every time the rate of generation of pollutant is doubled, the steady-state level will become doubled.

It is imaginable that the rate of removal of our substance may vary even more strongly with x than the first power, for example, that $\beta = c\,x^2$**. In this case the

---

* In chemical kinetics, if a reaction occurs that may be written $A + B \rightarrow C + D$, then the rate of disappearance of A, written $d[A]/dt$, equals $- k[A][B]$, where $[A]$ and $[B]$ are the concentrations of A and B, respectively, and k is a constant. This is the basis for writing that $\beta \propto x$.
** Some chemical reactions may satisfy an equation $A + A + B \rightarrow C + D$. For such a reaction the rate of disappearance of A satisfies an equation $d[A]/dt = - k[A]^2 [B]$. In our present notation, this sort of process would mark β as proportional to x^2.

concentration of the substance would reach an equilibrium level given by $x = (\alpha/c)^{1/2}$. For this case, if the generation rate α should abruptly be doubled to a new constant value, the ultimate equilibrium level would only be higher by $(2)^{1/2}$. Many variations of this sort are both possible and likely.

A complexity of the whole subject of atmospheric pollution is that both the formation and the destruction or removal of each pollutant occur by several different processes. The mathematical formulation shown here is nonetheless applicable. More terms may appear, but the basic form of the equations is unaltered. The basic fact that pollutants of natural origin have been evolving for eons yet exist in the atmosphere at a rather low and constant level demonstrates the existence of removal mechanisms operating in accordance with Equations (9)–(11) as opposed to Equations (7) and (8).

4. Photochemical Processes

The atmosphere absorbs a portion of the incident sunlight. It is well-known that sunburns are more severe at high altitudes than at sea level as a consequence of the absorption by the atmosphere. Part of the visible and infrared solar radiation is also absorbed by the atmosphere. These absorption processes are entirely separate from scattering and reflection by clouds and dust. The absorption of solar radiation may result in photochemical reactions in the atmosphere. The region from about 20 to 120 km altitude is often referred to as the chemosphere because of the chemical reactions between various atmospheric species. Some of these are photochemical reactions.

The basic store of energy in the atmosphere is in dissociated oxygen, produced during the day by radiation of 100 to 2450 Å. Averaged over the globe, O_2 is dissociated at the rate of 5×10^{11} cm^{-2} s^{-1}, by the process

$$O_2 + h\nu \rightarrow O + O. \tag{12}$$

Since the binding energy of O_2 is 5.1 eV, the photon energy $h\nu$ must be at least 5.1 eV or a wavelength shorter than 2450 Å. The ratio of O to O_2 increases with altitude beginning at about 20 km and exceeds unity at about 100 km altitude. Although most of the atmosphere is N_2, it has almost twice the dissociation energy of O_2 (9.76 eV). There are much fewer UV photons with 9.76 eV as compared to 5.1 eV. As a result there is little photodissociation of N_2 in the atmosphere. There is some photolysis of H_2O and O_3:

$$O_3 + h\nu \, (>0.9 \text{ eV}) \rightarrow O\,(^1D) + O_2\,(^1\Delta_g) \tag{13}$$

$$H_2O + h\nu \, (> 5.2 \text{ eV}) \rightarrow H + OH. \tag{14}$$

The net effect of the chemical reactions in the chemosphere is association of O atoms. The principal reactions are (Hunten, 1971):

$$O + O + M \rightarrow O_2 + M + 5.17 \text{ eV} \tag{15}$$

$$H + O_3 \rightarrow OH + O_2 + 3.34 \text{ eV} \qquad (16)$$

$$O + O_3 \rightarrow O_2 + O_2 + 4.06 \text{ eV} \qquad (17)$$

$$O + O_2 + M \rightarrow O_3 + M + 1.10 \text{ eV} \qquad (18)$$

$$OH + O \rightarrow O_2 + H + 0.72 \text{ eV}. \qquad (19)$$

Equation (18) is the primary means for the production of O_3. The maximum production occurs high in the chemosphere. The O_3 then migrates downward and is lost by reactions (16) and (17) and by reactions with SO_2 and with NO_2 if these contaminants are present. The maximum concentration of O_3, however, is found in the middle of the stratosphere at about 25 km (± 5) altitude as shown earlier in Figure 4. The location of the maximum concentration is the result of an equilibrium among various factors including decreasing density of O_2 with altitude, increasing UV radiation with altitude, migration of the O_3 downward, and increasing loss rate of O_3 downward. Ozone is an oxidant which is defined relative to any given substance as a chemical reagent capable of causing the given substance to change by reaction to a more electropositive or higher valent form. Ozone is a strong oxidant and causes such reactions in many substances. For example, $SO_2 + O_3 \rightarrow SO_3 + O_2$. The SO_2 is then said to have become oxidized to SO_3 and the O_3 is the oxidant or oxidizing agent. The O_3 is particularly active in oxidizing hydrocarbon contaminants in the atmosphere.

In a polluted atmosphere the photolysis of NO_2 can produce another oxidant, NO

$$NO_2 + h\nu \, (> 3.1 \text{ eV}) \rightarrow NO + O. \qquad (20)$$

NO can also be produced by the exothermic reaction

$$NO_2 + O \rightarrow NO + O_2 + 2.0 \text{ eV}. \qquad (21)$$

Thus, in a polluted atmosphere there are three oxidants than can react further with pollutants: O, O_3, and NO. This is why the Los Angeles type smog has been named "Photochemical Smog" for over 10 yr. The smog requires the action of sunlight to produce these oxidants photochemically. The details are still obscure and certainly very complicated. There are so many competing reactions with the various pollutants that one may never completely unravel the chemistry. The complexity can be seen in a review by Altshuller and Bufalini (1971). Many compounds, for example, nitric acid, nitrates, nitrites, nitrocompounds, aldehydes, ketones, peroxides, acyl-nitrates, and particles, absorb solar radiation resulting in the production of free radicals which in turn produce other new compounds (Cadle and Allen, 1970).

In general it is not clear as to which oxidant plays the key role in a given circumstance; for example, is the oxidant O or does O react to form O_3 by Equation (18) which is then the key reactant?

The NO may react by the inverse of Equation (20) or by

$$NO + O_3 \rightarrow NO_2 + O_2 + 2.0 \text{ eV}. \qquad (22)$$

The only way NO is really lost is by

$$NO + N \rightarrow N_2 + O + 3.3 \text{ eV}. \tag{23}$$

However, there is essentially no N in the lower atmosphere.

It has been suggested that singlet molecular oxygen, $O_2(^1\Sigma_g^+$ or $^1\Delta_g)$ may be particularly important in some reactions with pollutants. These are excited metastable states of O_2 with approximately 1.6 and 1.0 eV energy. The atomic oxygen oxidant $O(^3P)$ reacts rapidly with pollutants. It is estimated that $O_2(^1\Delta_g)$ might also have a high reaction rate and thus qualify as an oxidant. The reaction $O_3 + h\nu \rightarrow O_2 (^1\Delta_g) + O$ is very nearly a 'resonant process'. This term means that the energy of the photon needed is nearly zero. Exactly resonant processes occur spontaneously in either direction with high probability so that the change of O_3 to $O_2 (^1\Delta_g)$ occurs readily.

$O_2 (^1\Sigma_g^+$ or $^1\Delta_g)$ could be produced by the reactions in Equations (13) and (22). Photolysis of O_3 (Equation (13)) is probably the most important source, at least at the higher altitudes. They could also be produced by photoexcitation

$$O_2 (^3\Sigma_g) + h\nu \rightarrow O_2 (^1\Sigma_g^+) \quad \text{or} \quad O_2 (^1\Delta_g). \tag{24}$$

There could be energy exchange from O by

$$O (^1D) + O_2 (^3\Sigma_g) \rightarrow O (^3P) + O_2 (^1\Delta_g \quad \text{or} \quad ^1\Sigma_g). \tag{25}$$

Organic pollutants could absorb solar radiation and then transfer the energy to O_2 by

$$M + h\nu \rightarrow M^*$$
$$M^* + O_2 (^3\Sigma_g) \rightarrow M + O_2 (^1\Sigma_g \quad \text{or} \quad ^1\Delta_g). \tag{26}$$

Photochemical processes in the atmosphere do not require man-made pollutants. In Section 5 it is shown that the hydrocarbons (terpenes) from vegetation (especially pine trees) are much greater on a world-wide basis than from motor vehicles.

5. Types and Sources of Atmospheric Pollutants

It should be emphasized that there is no agreement on the quantity and type of pollutants. Some of the following are pollutants only under some conditions at some locations. Concentrations have extreme variability and depend on the source, topography, and meteorology. Table II lists some of the types and sources.

Much attention is given to the world-wide emission of a particular pollutant, and the quantity of it in the atmosphere is estimated. This approach has value in determining whether the quantity in Earth's atmosphere is increasing or decreasing. However, most of man's problems are local. If he is suffering from SO_2 it is of local origin. It is unimportant to him that most of the worldwide inventory of S in the atmosphere is from nature and the world-wide burden has not been increasing.

National yearly emissions of various man-made pollutants are popular statistics, especially in the United States. These large numbers tend to scare people, but it is difficult to find much use for them. Another popular presentation technique is to

TABLE II

Types and sources of atmospheric pollutants

Type	Source
CO_2	Volcanoes
	Burning fossil fuels
	Animals
CO	Internal combustion engine
	Volcanoes
Sulfur compounds	Bacteria
	Burning fossil fuels
	Volcanoes
	Sea spray
Hydrocarbons	Internal combustion engine
	Bacteria
	Plants
Nitrogen compounds	Bacteria
	Combustion
Particles	Volcanoes
	Wind action
	Combustion
	Industrial processing
	Meteors
	Sea spray
	Forest fires

quote the percentage of a particular pollutant from each of several sources, such as that automobiles produce one-half of the hydrocarbons in the United States. Since most atmospheric pollutants are a local problem, there is no locality which is the national average. The most useful information is the sources and amounts of each pollutant in a limited region, considering topography and meterology. The approp-

TABLE III

Sources of particulates in New York City

November 1969

Source	%	kg/yr
Space heating (apartments)	32.3	2.03×10^7
Municipal incineration	19.3	1.22×10^7
Apartment incineration	18.4	1.16×10^7
Mobile source	14.3	0.90×10^7
Power generation	9.2	0.58×10^7
Industrial	6.5	0.41×10^7
Total		6.3×10^7

riate data base is different upwind than down wind. With caution, a listing of sources and quantities for a region such as New York City is useful such as that for particulates shown in Table III (Eisenbud, 1970). This clearly indicates that incineration and apartment heating are the source of almost three-fourths of the soot and dirt in New York City.

There should be national and international standards for air quality. There is nothing illogical about having the same minimum standards at all locations. This is not synonymous with having the same identical solution for all regions. To achieve air quality standards in Los Angeles, it seems obvious that improvements in motor vehicle emissions must be made. However, it can be argued that it is a waste of national resources to require automobiles in many other regions, such as North Dakota, to comply with specifications needed for Los Angeles, or some other metropolitan area.

While national statistics should not serve as the basis for specific local action, they can focus attention on potential prominent sources of pollutants. Approximately 90% of the mass of all pollutants is emitted as a gas compared to 10% as particles

TABLE IV

Natural vs. man-made pollutant sources

kg yr^{-1}

	Natural	Man-made
O_3	1.8×10^{12}	small
CO_2	7.2×10^{13}	1.4×10^{13}
H_2O	4.5×10^{17}	$9 \ \times 10^{12}$
CO	?	1.8×10^{11}
S	1.3×10^{11}	6.8×10^{10}
N	1.4×10^{12}	1.8×10^{10}

and liquids. Table IV (Robinson and Robbins, 1968, 1969) provides an estimate of the yearly source of some pollutants from natural and man-made sources. All of these estimates are extremely difficult to make.

Table V shows the distribution of world-wide S sources (Robinson and Robbins,

TABLE V

World-wide sulfur emissions

Source	kg yr^{-1}
SO_2 – Coal combustion	4.7×10^{10}
SO_2 – Petroleum combustion	1.3×10^{10}
SO_2 – Smelting	$7 \ \times 10^9$
H_2S – Bacteria	8.8×10^{10}
SO_4 – Sea spray	4.0×10^{10}
Total	19.5×10^{10}

1970b) and that two-thirds of the S comes from natural sources. They point out that two-thirds of the total S is emitted in the Northern Hemisphere. This results from the fact that over 90% of the man-made S is emitted in the Northern Hemisphere.

Essentially all of the N emitted into the atmosphere comes from bacteria as shown in Table VI (Robinson and Robbins, 1970a).

TABLE VI

World-wide nitrogen emissions

Source	kg yr^{-1}
NH$_3$ – Bacteria	8.6×10^{11}
N$_2$O – Bacteria	3.4×10^{11}
NO – Bacteria	2.1×10^{11}
NO$_2$ – Coal combustion	7.4×10^9
NO$_2$ – Space heating	4.5×10^9
NO$_2$ – Motor vehicles	2.0×10^9
NO$_2$ – Other combustion	0.7×10^9
NH$_3$ – Combustion	3.1×10^9
Total	14.28×10^{11}

Aerosols can be emitted directly into the atmosphere and can be formed in the atmosphere from gaseous products. Most of the world-wide aerosol content is produced from natural sources as shown in Table VII (Rosen, 1969; Robinson and Robbins, 1971). Although the natural sources dominate, in our urban region the man-made aerosols may be much more important. Although the average volcanic source is small, a magnitude 1 volcano may inject more particles at one time than all other sources combined.

TABLE VII

World-wide aerosol emissions

Source	kg yr^{-1}
Sea salt	9×10^{11}
Natural H$_2$S	1.8×10^{11}
Natural N compounds	6.3×10^{11}
Natural terpenes	1.8×10^{11}
Man-made particles	8.3×10^{10}
Man-made SO$_2$	1.3×10^{11}
Man-made N compounds	2.7×10^{10}
Man-made hydrocarbons	2.5×10^{10}
Wind dust	1.8×10^{11}
Forest fires	2.7×10^9
Volcanoes (average)	3.6×10^9
Meteoric dust	5×10^9
Total	23.5×10^{11}

The estimates of the national yearly quantities of pollutants for several different sources are shown in Table VIII. The numbers should not be taken too seriously; but they do indicate major sources. Data indicate that CO, sulfur compounds, nitrogen compounds, particulate matter, and hydrocarbons are not accumulating in earth's atmosphere and are a local problem (USDHEW, 1966).

TABLE VIII

Yearly general air pollution sources in the U.S.A.

Units 10^9 kg

Source/Type	S–O's	N–O's	CO	Particulates	Hydrocarbons
Electric power	12	4	1	4	1
Space heating	6	1	2	1	1
Motor vehicles	1	6	60	1	10
Industry	10	2	2	6	4
Refuse disposal	1	1	1	1	1

Much concern has been expressed over potential environmental effects of the operation of a large number of SST's. Air environmental considerations require analysis of the effects of the exhaust products from the SST. It has been suggested that these products may result in air pollution, modification of the weather, and a change in the Earth-Sun heat balance.

Before any of these can be investigated, the first important step is to study the local atmospheric chemistry effects when SST exhaust products are dumped into the stratosphere. The partially ionized exhaust products may react in the atmosphere under both sunlit and non-sunlit conditions. The problems are so complex that in practice one may have to monitor the effects of SST exhaust products to determine their importance.

The SST will operate in the stratosphere at 16 to 22 km altitude. The exhaust products will depend on engine size, design, fuel, and operating conditions. A preliminary estimate of the injection of exhaust products into the stratosphere can be made.

Assume that each SST has three engines burning a total of 20 kg s^{-1} of fuel. If the fuel is $C_{10}H_{22}$ (molecular weight 142 g mole^{-1}), then each SST burns 140 mole s^{-1}. The exhaust products are then

CO_2 62 kg s^{-1} (1400 moles s^{-1})
H_2O 28 kg s^{-1} (1540 moles s^{-1}).

If the jet flies at Mach 1.7, it moves about 500 m s^{-1} and thus deposits 124 g of CO_2 per linear meter and 56 g m^{-1} of H_2O. At about 20 km altitude, the temperature is ~216 K. At this temperature saturation vapor pressure of water amounts to slightly over 0.02 g m^{-3}. Since it is notable that jets do NOT form contrails at this altitude, the 56 g m^{-1} of H$_2$O must attenuate over more than (56 g m^{-1})/(0.02 g m^{-3}) = 2800 m^2 or more than 30 m radius. The CO$_2$ is doubtless similarly attenuated.

A figure sometimes mentioned is 500 SST's flying 12 hr per day. Thus the quantities of CO$_2$ and H$_2$O s^{-1} need to be multiplied by 7.7×10^8 to obtain total quantities yr^{-1}. There are small quantities of other exhaust products as indicated in Table IX.

TABLE IX

SST emission products per year

Product	kg
CO$_2$	4.8×10^{10}
H$_2$O	2.2×10^{10}
CO	2.5×10^9
NO	2.5×10^9
SO$_2$	5×10^7
Particles	5×10^7
Hydrocarbons	5×10^6

The SST's main contribution to the environment would be CO$_2$; however, the atmosphere contains 1.7×10^{15} kg of CO$_2$. In addition, the burning of fossil fuels is much larger than the SST contribution. H$_2$O is the other main contributor.

The amounts of CO and hydrocarbons are quite insignificant and NO is small compared to the natural levels of NO$_2$. It has been argued that the exhaust particles and particles formed on SO$_2$ may affect the stratospheric aerosol concentration. Meteoric dust deposits an estimated 10^9 to 10^{10} kg yr^{-1} into the atmosphere. There is also a sizeable aerosol contribution into the stratosphere from the troposphere. Volcanoes have injected as much as 10^{13} kg of dust in a single eruption into the stratosphere without catastrophic effects. It does not seem that any argument can be made against particulate matter from SST's.

It has been suggested that the deposit of H$_2$O in the stratosphere by SST flights would lead to the destruction of the O$_3$ layer which absorbs a significant amount of the solar UV, as shown by the photolysis reaction in Equation (13). The O$_3$ absorption maximum is at 2700 Å with a half maximum width about 500 Å. The O$_3$ concentrations with altitude shown in Figure 4 are the result of various production and loss mechanisms. These average O$_3$ concentrations show a variation in the stratosphere of an order of magnitude. The time to time and place to place variations are even more spread.

The SST would deposit about 2×10^{10} kg yr^{-1} of H$_2$O, about 1 % of the amount injected into the stratosphere by natural means. The natural H$_2$O is believed to be mostly injected into the stratosphere at mid-latitudes. There must be a large variation in the injection of natural H$_2$O into the stratosphere as a function of time and space. It may be difficult to measure the SST deposit of H$_2$O.

The hydrogen-ozone chemistry above the tropopause is very complex. There are a large number of potential reactions and excited states (Hunt, 1966; Gattinger, 1971).

There are suggestions that OH which can be produced from H_2O will act to remove O_3. While it is quite true at altitudes above 60 to 70 km, it is not necessarily correct in the stratosphere, where it is also not apparent that the amount of H_2O is a limiting factor. Reaction rates, species, and concentrations above the tropopause are poorly known; however, all observations to date indicate that the concentrations of various species are extremely variable.

6. Sulfur and Sulfur Compounds

In point of inconvenience, annoyance, toxicity, and actual damage, S and its compounds have a high place on any list. In fact, the damaging effects are so severe that the S compounds rate a top place despite lower actual concentrations than some other common pollutants. Sulfur is found on the Earth in both elemental form and as compounds, including more complex forms such as proteins. It constitutes about 0.06% of the Earth's mass and approximately the same fraction of the universe. Sulfur exists in the atmosphere largely in one of three forms: H_2S, SO_2, and SO_4 compounds, particularly $(NH_4)_2 SO_4$ and H_2SO_4 (see Table V).

The sulfur cycle in the environment is very complex and many aspects are poorly understood (Robinson and Robbins, 1970b). Hydrogen disulfide (H_2S) is evolved in some organic processes and constitutes over 45% of the S emitted to atmosphere. Next is sulfur dioxide (SO_2) which is largely man-made and constitutes one-third of the total S emitted. Most of the SO_2 is emitted by burning fossil fuels containing S. The other source is SO_4 aerosols from sea spray.

It is estimated that two-thirds of the S is emitted in the Northern Hemisphere and this results mostly from the fact that over 90% of the SO_2 pollution is in the Northern Hemisphere. Studies in glacier ice indicate that in spite of the increasing industrial burning of fuels emitting SO_2 the world-wide background has not increased. The sinks of S are not understood. The ocean alone contains about 10^{18} kg of S and is believed to be an important sink for S.

Sulfur thus enters the atmosphere from natural sources at a rate that up to the present has exceeded the output from human sources. Volcanic sources sustain a continuous level of emission with enormous peaks superimposed during major eruptions. Since 1963, an average of one eruption per year has injected volcanic matter to heights above the tropopause. Sea spray may inject even more than volcanoes. The intense concentrations appear to diminish rapidly with distance, even from continuous sources, so that the world-wide level of SO_2 appears to be at 1 to 10 ppb. It seems now to be established that pollutants at altitudes above the tropopause become widely distributed, even to the extent of crossing the equatorial barrier that effectively separates the hemispheres at lower levels.

The total amount of S estimated to be in the atmosphere is a number that is of the same order of magnitude as the estimated amount injected into the atmosphere from all sources in a year. Because of the large percentage that occurs naturally and has been doing so over geological time spans, it is clear that an equilibrium between addition and removal exists following some mathematical formulation such as that shown in Section 3. The quantity of S in the total atmosphere is placed in the range of 10^{11} kg.

A. H₂S CHEMISTRY

The emission of H_2S into the atmosphere, both naturally and by man is completely dominated by bacterial emission. There are some industrial and volcanic emissions which could be a local pollutant. Hydrogen sulfide is used extensively in the chemical industry, certain plants being enveloped for a distance of several km with its powerful aroma. It is extremely toxic, highly flammable, and corrosive to the eyes and respiratory tissues. Maximum allowable concentration is set at 20 ppm. In high concentrations, it can cause death quickly, and at such levels, it may be unidentified because it paralyzes the olfactory nerves.

It appears that H_2S is oxidized in the atmosphere to SO_2 or is utilized by certain bacteria in photosynthesis. The oxidation process involves O_3 in a complicated reaction on the surface of aerosol particles. Depending on the availability of oxidant the lifetime of H_2S in the atmosphere is estimated to be a few hours to days. The natural H_2S background in the troposphere is estimated to be $\sim$0.2 ppb.

B. SO₂ CHEMISTRY

When sulfur burns, SO_2 is formed, and this, in addition to being noxious is intensely smog-forming and possibly the worst 'problem' gas in the atmosphere. The Handbook of Dangerous Materials (Sax, 1963) declares 10 ppm to be the maximum allowable concentration, 0.3 to 1 ppm to be detectable by human senses, 50 to 100 ppm to be the maximum allowable for periods of 0.5 to 1 hr, and 400 to 500 ppm to be immediately dangerous to life. A concentration of less than 1 ppm is believed to be injurious to foliage. Concentrations as low as 0.1 ppm continuously are believed to be injurious to individuals with lung problems. Fortunately, it is so highly irritating to mucous tissues that it provides its own warning of toxic concentrations.

The most widespread human source of SO_2 is from the burning of hydrocarbons that contain S as an impurity, most notably coal. The concentration of SO_2 in cities, of which New York may be the worst, reaches 1 and even 1.5 ppm under 'normally bad' conditions, for example, in wintertime on days with still air and a low-lying inversion layer.

H_2S can burn in air, the reaction being

$$2\,H_2S + 3\,O_2 \rightarrow 2\,H_2O + 2\,SO_2 + 15\ \text{kcal}\ (0.65\ \text{eV}).\qquad (27)$$

Since the reaction is exothermic, it can occur without photochemical action. Whether this reaction is the main one responsible for converting H_2S to SO_2 does not appear to be established although it is certain that the H_2S is largely converted to SO_2. If either O or O_3 is present, it would also contribute to the conversion but reaction rates do not appear to be known. The common uses of H_2S in chemical processes (for example to produce the difficultly soluble metallic sulfides) do not seem to enter into atmospheric chemistry.

When H_2S burns in insufficient O_2 the newly formed SO_2 reacts with the excess H_2S by the reaction

$$2\,H_2S + SO_2 \rightarrow 3\,S + 2\,H_2O \tag{28}$$

and elemental sulfur is produced. The production of free S during a burning process seems at first to be surprising because free S burns readily in O_2 to form SO_2. The shortage of O_2 is the key. The process may occur during volcanic eruptions as the emission of S is believed to have been observed.

Most of the SO_2 is ultimately converted to sulfates in aerosols and precipitates out of the atmosphere. Some SO_2 is also washed out by rain and absorbed by vegetation and water. The natural tropospheric background of SO_2 is ~ 0.2 ppb while the amount of sulfur tied up in aerosol sulfates is $7 \times 10^{-7}\ \mu g\ cm^{-3}$ (or roughly 0.3 ppb).

The details of what becomes of SO_2 in the air are both complex and uncertain. The simplest possibilities are (1) reaction with atmospheric water

$$SO_2 + H_2O \rightarrow H_2SO_3 + 18\ \text{kcal}\,(0.78\ \text{eV}) \tag{29}$$

and (2) oxidation by atmospheric O_2

$$SO_2 + \tfrac{1}{2}O_2 \rightarrow SO_3 + 22\ \text{kcal}\,(0.95\ \text{eV}). \tag{30}$$

Both reactions are possible but appear to be slow. The H_2SO_3 produced in reaction (29) is unstable and changes spontaneously in air to H_2SO_4. The SO_3 produced by reaction (30) is extremely hygroscopic, the reaction involved being

$$SO_3 + H_2O \rightarrow H_2SO_4 + 38\ \text{kcal}\,(1.65\ \text{eV}). \tag{31}$$

This is a large amount of heat for a hydration process. The resulting H_2SO_4 is itself strongly hygroscopic, and the molecule rapidly grows into a droplet. Sulfuric acid mist is well known in chemical laboratories. In the atmosphere, it may well be one of the constituents of smog.

There are other processes whereby SO_2 can be converted into H_2SO_4. Thus if NO_2 and water are present, there is a reaction

$$SO_2 + NO_2 + H_2O \rightarrow H_2SO_4 + NO. \tag{32}$$

This reaction is employed in the manufacture of H_2SO_4 by 'the lead chamber' process, but it is only recently that the work by Bricard (1971) has disclosed the importance of this reaction at concentrations of SO_2 and of NO_2 at such low levels as 0.5 ppm as a result of the action of sunlight (see Section 7). Since the sulfuric acid in turn forms mist, it is of interest to note that reaction (32) produces a full 100 times as much mist as reactions (29) and (30) without NO_2.

The reaction

$$SO_2 + O + M \rightarrow SO_3 + M + 81\ \text{kcal}\,(3.6\ \text{eV}) \tag{33}$$

becomes important if any atomic oxygen is present. In Section 7, it is shown that the action of sunlight on NO_2 liberates O so that this reaction doubtless contributes significantly to photochemical smog.

Some oxidation of SO_2 to SO_3 is doubtless produced by hydrocarbons and will be mentioned under that heading.

Once H_2SO_4 has been formed, the sulfate radical can combine with the baser metals and NH_3 to produce solid sulfates. The existence of SO_4 at stratospheric levels is known although it is more a matter of conjecture that it is in the form of H_2SO_4 and $(NH_4)_2SO_4$. The extreme insolubility of compounds like $BaSO_4$ makes this or similar compounds possible end products except for the rarity of Ba, Ca, and similar ions in the atmosphere. Very recent observations have disclosed many metals including Ba, Ca, Pb, Hg, and others in single droplets of smoggy atmosphere containing largely H_2SO_4.

The presence of increasing concentrations of H_2SO_4 in the lower atmosphere is indicated by the increasing and more widespread occurrence of 'acid rain' (American Chemical Society, 1969). In Europe, it has spread northward and values of pH down to 4 are noted in Sweden. In Pennsylvania, rain with pH of 3 and even 2 has been reported. This sort of acidity in fresh water lakes has severe ecological consequences. It would appear that the problem of acid rain is growing from a purely local to a regional one.

If the rate of removal of SO_2 follows the simplest possible law treated in Section 3 according to which the removal rate is simply proportional to the concentration, then doubling the rate of production doubles the equilibrium concentration level. Since the global level is in the range of ppb, concerns about global levels in the catastrophic range can only be threatened when the world-wide production rate of SO_2 increases by a factor of 100 to 1000.

In the study of global concentrations the factor of wind dissipation of course, has no significance except perhaps as to the speed with which a steady state is reached. At the regional level, as well as the local, the details of the processes of dissipation and dilution influence the effective levels that exist. At least one piece of evidence exists concerning regional levels of SO_2 concentration, although a great deal more is needed. In 1969–70, an SO_2 detector was installed on a ship plying between London and Gothenburg (Brosset and Marsch, 1970). At a time when the SO_2 concentration over London ranged from 1.65 to 1.83×10^{-4} μg cm^{-3}, by the time the ship was 50 km down the Thames and despite the fact that the wind direction was also down the Thames from London, the SO_2 reading was down to 9×10^{-6} μg cm^{-3} and range from 3 to 9 as the ship crossed the North Sea.* It is thus necessary to conclude that even with as large and spread out a source of SO_2 as the city of London has, even down wind the SO_2 has vanished in a distance of 50 km.

Efforts at removing SO_2 from stack gases follow several basic lines: (1) scrubbing, i.e., dissolving SO_2 in water, (2) precipitation as an insoluble sulfate by injection of

---

* Since 1 cm^3 of air weighs 1.2×10^{-3} g and SO_2 has a molecular weight 2.2 times that of air, a concentration of 1.9×10^{-6} μg cm^{-3} of SO_2 is equivalent to 1 ppb of SO_2 at NTP as a molecular count.

limestone dust, and (3) catalytic oxidation of SO_2 to SO_3. In these various processes, H_2SO_4 and free S are commonly by-products. The total S effluent of all utilities could supply half of the present S market.

7. Ammonia and the Oxides of Nitrogen

There are many known combinations of oxygen and nitrogen. Since they are toxic to human beings in varied ways, it must be regarded as fortunate for human existence that nitrogen and oxygen are so resistant to spontaneous reaction. Elevated temperatures, electric discharges, or indirect reactions are required to induce their reaction.

Four nitrogenous compounds are found in the natural atmosphere. Of these N_2O (nitrous oxide) is the most plentiful, with a global concentration of 0.25 ppm. As with most natural nitrogen compounds, N_2O arises from biological processes, probably from bacteria in the soil and appears to take no part in atmospheric chemistry or pollution.

Ammonia is also synthesized by many biological processes and is found at a worldwide level of 6 ppb in the atmosphere. Its concentration is probably as low as this because of removal by SO_4 as solid $(NH_4)_2 SO_4$. Except for this role in removal of SO_4, NH_3 is probably relatively uninvolved in air pollution chemistry although at local levels it is evolved in the burning of coal, present from the vegetable origin of the coal. In fact a commercial source of NH_3 has been as a by-product from coke ovens (Latimer and Hildebrand, 1929).

A. NO/NO₂ CHEMISTRY

The remaining oxygen compounds of significance are NO and NO_2. Nitrogen dioxide forms a 'double' molecule or *dimer* N_2O_4 if it is at all concentrated and while distinguishable from NO_2 has almost identical properties. At elevated temperatures, e.g., 100°C, it dissociates back to NO_2. In unpolluted air both NO and NO_2 are only present in abundances measured in ppb. They may be produced biologically, by lightning, and in the higher atmosphere in very small quantities by high energy radiations. The background levels are shown in Table X (Robinson and Robbins, 1970a). Table X also shows NO_3 aerosols and NH_4 aerosols, secondary substances that are discussed later.

TABLE X

Nitrogen compound background

Compound	Concentration
NO	0.2 to 2 ppb
NO₂	0.5 to 4 ppb
NO₃ aerosols	2×10^{-7} μg cm^{-3}
NH₄ aerosols	1×10^{-6} μg cm^{-3}

Both NO and NO_2 are, however, important components of the polluted atmosphere and require considerable attention in studies of local contamination. Nitric oxide is produced in hot fire boxes, furnaces, and particularly in modern automobile engines operating at high combustion temperatures. Typically, a modern (1970) American car shows exhaust concentrations of NO around 2800 ppm as compared with values of 500 ppm for older cars operating at lower cylinder temperatures.

The basic reactions and resulting oxides of interest in pollution concerns are:

Formation of NO

$$\tfrac{1}{2}N_2 + \tfrac{1}{2}O_2 \rightarrow NO - 21.6 \text{ kcal} (0.935 \text{ eV}) \tag{34}$$

Formation of NO_2

$$\tfrac{1}{2}N_2 + O_2 \rightarrow NO_2 - 8.3 \text{ kcal} (0.36 \text{ eV}) \tag{35}$$

Conversion of NO to NO_2

$$NO + \tfrac{1}{2}O_2 \rightarrow NO_2 + 14.2 \text{ kcal} (0.62 \text{ eV}). \tag{36}$$

It is to be noted that neither reactions (34) nor (35) occur spontaneously at room temperature, but that (36) does. We shall be pointing out various atmospheric phenomena regarding these oxides. The values of the heats of reaction shown here are for 288 K and they change with temperature. Reaction (36), for example, is quite strongly reversed in direction by 600 °C. Because of this reversal of reaction (36), at elevated temperature, hot fireboxes and engines produce NO rather than NO_2.

Some preliminary characteristics of the oxides may be noted. Nitric oxide is colorless, and compared with other oxides of nitrogen, has a very low boiling point ($-$ 152 °C). It is not an irritant. In human beings it may react with hemoglobin, and in animals it has been found to attack the central nervous system. It may be regarded as not dangerous in its own right, in air, because of reaction (36) above. By this reaction, NO at potentially dangerous concentrations reacts with air changing to NO_2. The conversion of NO to NO_2 has been of interest to chemical kineticists for a full 50 yr as one of the few clear cut examples of a termolecular reaction. Written from the kinetic point of view, the reaction is

$$NO + NO + O_2 \rightarrow 2 NO_2. \tag{37}$$

The rate of disappearance of NO may thus be written

$$\frac{d[NO]}{dt} = - k[NO]^2 [O_2] \tag{38}$$

where brackets around a chemical symbol denote the concentration of that compound, expressible in molecules cm^{-3} or moles l^{-1} or other appropriate units. The reaction rate is seen to depend on the square of the concentration of NO, a feature that is effective in preventing the NO level in air from remaining at a high level for any length of time but makes the reaction quite slow at levels of a few ppm of NO.

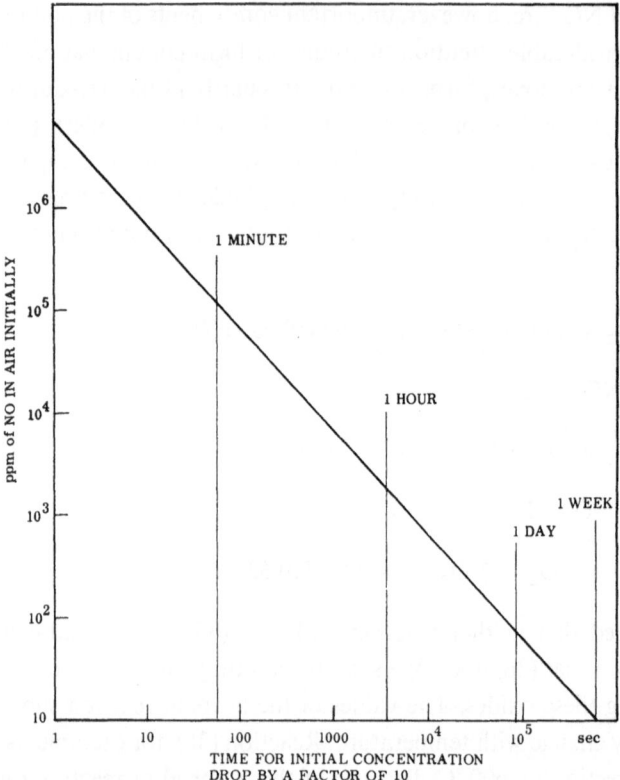

Fig. 7.   Rate of the reaction $2NO + O_2 \rightarrow 2NO_2$, the spontaneous oxidation in air of NO to NO_2. A rate constant of 10^{-38} cm^2 s^{-1} is used. The graph shows that an initial concentration of 10 ppm would diminish to 1 ppm in one week.

Figure 7 is a log-log plot showing what occurs if a burst of NO is admitted to air at a concentration shown on the ordinate scale. The time after admission is plotted on the abscissa and the curve discloses the time for the concentration to drop from the indicated ordinate value to one-tenth of the initial value. Concentrations in excess of 10^3 ppm thus fall fairly rapidly. By contrast, concentrations of 1 to 10 ppm (not shown in the curve) are scarcely altered by the conversion reaction.

Nitrogen dioxide has little resemblance to NO. It is a red-brown gas with a high boiling point of 21 °C. The gas is extremely irritating to mucous tissues. At lower temperatures (below 100 °C) it is largely the double or dimerized molecule N_2O_4 with a more orange color but essentially the same toxic action.

The NO in the various exhaust gases, whether from stacks or tail pipes, changes rather rapidly after emergence into cool surroundings into NO_2 (or N_2O_4). A brownish plume is often visible over the stacks of electric plants even though they are not coal burning. The brown haze that develops near the ground on still-air days, notably near freeways, is largely NO_2.

Nitric oxide is oxidized to NO_2 much more rapidly by O_3 than by O_2. The reaction rate for

$$O_3 + NO \rightarrow NO_2 + O_2 + 48.5 \text{ kcal } (2.10 \text{ eV}) \tag{39}$$

has been measured, and the rate constant is $5 \times 10^{-14} \text{ cm}^3 \text{ moles}^{-1} \text{ s}^{-1}$. That reaction (39) is much faster than (37) is influenced by the fact that (37) is a three-body reaction requiring two NO molecules at once, whereas (39) requires only a single NO molecule. Since NO is only being considered in this context in concentrations of a few ppm at most, the speed of (39) relative to (37) is the more marked.

In sunlight, NO_2 is dissociated, the reaction being

$$NO_2 + h\nu \, (> 3.12 \text{ eV}) \rightarrow NO + O. \tag{40}$$

This reaction is of great importance in photochemical smog because of the production of O at much lower altitudes than is possible by photodissociation of O_2. The O at these altitudes can produce O_3 by

$$O + O_2 + M \rightarrow O_3 + M + 24.2 \text{ kcal } (1.05 \text{ eV}) \tag{41}$$

and SO_3 by

$$SO_2 + O + M \rightarrow SO_3 + M + 83 \text{ kcal } (3.6 \text{ eV}). \tag{42}$$

Reaction (40) has been experimentally verified by Bricard (1971) and enhances the rate of formation of H_2SO_4 (see reaction (32) in Section 6), when NO_2 is present. O_3, as well as O, is important in hydrocarbon processes (see Section 9). Finally, any traces of O_3 and O that do not contribute to smog can react to restore NO to NO_2 which thereby remains available for further photochemical dissociation.

Both NO_2 and N_2O_4 react with water to form HNO_3 and also HNO_2. The principal reaction involved when only small concentrations of NO_2 are available is:

$$2 NO_2 + H_2O \rightarrow HNO_2 + HNO_3. \tag{43}$$

Nitrous acid is unstable with respect to decomposition by the reaction

$$3 HNO_2 \rightarrow HNO_3 + 2 NO + H_2O \tag{44}$$

but this reaction is both slow and readily reversible. The presence of HNO_3 in the atmosphere is severely irritating particularly to mucous tissues. The net result of these two reactions is

$$3 NO_2 + H_2O \rightleftarrows 2 HNO_3 + NO. \tag{45}$$

This reaction (Cadle and Allen, 1970) acts in humid air at room temperature to convert about 5% of the NO_2 to HNO_3.

Ammonia is extremely soluble in water so that any traces of it are likely to form NH_4OH. While NH_4OH is a weak base and only dissociates partially into NH_4^+ and OH^-, the existence of NH_4^+ is significant, for example, in the formation of $(NH_4)_2SO_4$. Whether a droplet of concentrated H_2SO_4 has an actually greater attraction for NH_3 than a droplet of pure water is not known and would be of interest as a step toward the formation of $(NH_4)_2SO_4$.

The role of nitrogen compounds is thus of great importance in the polluted atmosphere, but even combined with SO_2, it is inadequate to explain all of the smog phenomena of the Los Angeles area. For this, the role of hydrocarbons must be studied.

8. Carbon Dioxide and Carbon Monoxide

Carbon dioxide is the least abundant of the four basic constituents of the dry atmosphere, its concentration being approximately 320 ppm in 1970. Far from being toxic to man, this compound stimulates the breathing nerve without which action man would not breathe subconsciously. It is obviously suffocating if it becomes so concentrated as to deplete the oxygen supply. As the concentration is raised above the normal level, the rate of breathing becomes increased, 3×10^4 ppm doubling the rate. At higher concentrations still, the respiratory center becomes paralyzed, 10^5 ppm being endurable for only a few minutes.

The balance of CO_2 in the atmosphere is the result of many different processes of absorption and generation, many of which need further study (Johnson, 1970).

Volcanic eruptions are an unknown source of CO_2; however, the 1963 and 1965 major eruptions did not seem to affect the quantity of CO_2 world wide. It has been estimated that on the average the CO_2 from volcanoes in the last 50 yr has only been 1 to 2% of that emitted from fossil fuel burning in the same period. Although volcanic CO_2 was dumped suddenly into the atmosphere, no temperature effects or changes in CO_2 concentration could be detected. The burning of carbon compounds by animals and bacteria, the reverse of the photosynthesis process, produces 80% of the CO_2 ($\sim 7.2 \times 10^{13}$ kg yr^{-1}):

$$CH_2O + H_2O + O_2 \rightarrow CO_2 + 2H_2O + energy \quad \text{or}$$
$$CH_2O + H_2O + 2S \rightarrow CO_2 + 2H_2S + energy. \tag{46}$$

The next major source of CO_2 is the burning of fossil fuels which produces about 20% of the total ($\sim 1.4 \times 10^{13}$ kg yr^{-1}). The amount of CO_2 in the atmosphere is increasing by about 1.2 ppm per year. About 0.2 of the CO_2 being emitted by the burning fossil fuels is remaining in the atmosphere. Allowing for an accelerating rate of fuel burning, the SCEP report (1970) projects 379 ppm of CO_2 in the atmosphere by the year 2000. Figure 8 shows the concentration of CO_2 in the atmosphere for the last 100 years and indicates the predicted concentration.

The sinks for the CO_2 cycle are poorly understood (Bolin, 1970); however, plants use CO_2 in photosynthesis. It is well known that plant growth is increased as the CO_2 concentration increases. Part of the CO_2 must be absorbed into the oceans; however, calculations do not quantitatively remove the necessary CO_2 from the atmosphere to explain observations.

Carbon dioxide dissolves in water, with the process reaching a limit when the value of pH becomes approximately 5. The process actually is:

$$CO_2 + H_2O \rightarrow H_2CO_3 \rightarrow H^+ + HCO_3^-. \tag{47}$$

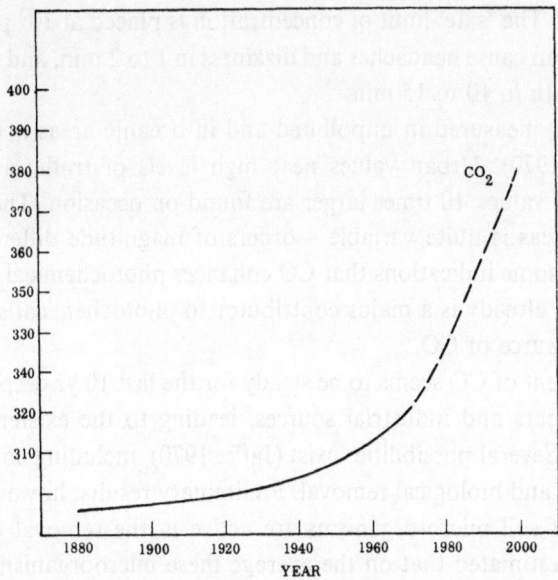

Fig. 8. Concentration of CO_2 in the atmosphere for the last 100 years.

Even distilled water that has been condensed in atmospheric air is acid to this extent. The splashing and foaming of waves, the falling of rain drops and waterfalls, all act to remove CO_2 from the air. The water would become saturated unless further reactions continually removed CO_2. Such processes may include oceanic floral growth including plankton, and precipitation of $CaCO_3$ as coral or other carbonates. Some CO_2 is converted into limestone, $Ca(HCO_3)_2$.

It was suggested for a number of years that the increase in CO_2 in the atmosphere would cause a rise in the mean temperature of the world. A rise of 5 °C of the normal ground temperature could melt the polar ice caps. Statistics appear to disclose that the mean temperature of the atmosphere rose 4 °C between 1840 and 1940 with a few tenths of a degree drop since 1940. Therefore, although the atmospheric concentration of CO_2 has steadily risen the temperature at the Earth's surface has started to decrease. Thus, we do not understand the processes affecting the surface temperature.

Carbon monoxide exists chiefly as the result of incomplete oxidation of carbons and hydrocarbons. It is estimated that $\sim 2 \times 10^{11}$ kg yr^{-1} of CO are produced. Most of the source is man made. The internal combustion engine is the largest single source contributing about two-thirds of the total. The smoking of cigarettes is estimated to give 4×10^4 ppm in the smoke and that smoking one package per day is equivalent to a constant environment of 5×10^1 ppm. Some CO is produced in forest fires and there are a number of biological sources in the sea; however, they have not been well studied. In marked contrast to CO_2, it is toxic to human beings. Its toxic action appears to be that after inhalation, it reacts with hemoglobin in the blood forming a stable compound that no longer reacts with oxygen. The hemoglobin thereby loses its power to transport oxygen from the lungs to the cells, and the victim is threatened with asphyxiation. The 'safe' limit of concentration is placed at 10^2 ppm while a level of 6.4×10^3 ppm can cause headaches and dizziness in 1 to 2 min, and unconsciousness and hazard of death in 10 to 15 min.

The level of CO measured in unpolluted and in oceanic areas is (1969–70) 0.1 to 0.2 ppm (SCEP, 1970). Urban values near high levels of traffic average at about 10 to 20 ppm and values 10 times larger are found on occasion. The distribution of CO in polluted areas is quite variable – orders of magnitude difference with a few meters. There are some indications that CO enhances photochemical smog problems. The motor vehicle already is a major contributor to photochemical smog and is also the major single source of CO.

The global content of CO seems to be steady for the last 10 yr despite the increased production from cars and industrial sources, leading to the existence of an as yet unidentified sink. Several possibilities exist (Jaffe, 1970), including conversion to CO_2 in the atmosphere and biological removal. Preliminary results, however, indicate that a large number of soil micro-organisms are active in the removal of CO from the atmosphere. It is estimated that on the average these microorganisms could remove twice as much CO per unit time as is produced. In certain urban regions, this loss mechanism is over burdened and CO levels can build up. No atmospheric reaction for converting CO to CO_2 has been found that is rapid enough to keep the global concentration constant. The ocean is probably a source of CO.

9. Hydrocarbons

Hydrocarbons range from simple marsh gas (CH_4) evolving from decaying organic material to an almost unlimited degree of complex compounds. Under this heading many chemical sprays like DDT may be included, some highly toxic to human beings, some benign, some beneficial. Man probably only produces about 15% of the global hydrocarbons emitted, but in urban areas, where the hydrocarbons participate in photochemical smog, the man-made sources dominate. The natural sources are CH_4 from bacterial decomposition and the terpenes from forests and vegetation. It has been estimated that over 10^{11} kg yr^{-1} of hydrocarbons are emitted by forests and vegetation (see Table VII). These cause the blue haze seen over the Allegheny Mountains.

The usual and obvious precautions for protection against dangerous, local concentrations are necessary. Only a few hydrocarbons ever reach dangerous regional intensity levels, and even in these cases, the hazard is associated with the secondary formation of photochemical smog rather than with primary toxicity. Such problems are correlated with large numbers of motor vehicles in limited areas like the Los Angeles basin with added complexity due to stagnant air and a low-lying inversion layer.

While the hydrocarbons associated with the refining, burning, and evaporation of gasoline and the motor fuels or natural hydrocarbons are essential ingredients for the formation of photochemical smog, they alone cannot account for severe smog without the associated oxides of nitrogen and probably oxides of sulfur and carbon. Present evidence is that oxidants are a key to smog and that O, O_2, O_3, NO, and NO_2 are not sufficiently abundant to constitute the observed oxidants without the addition of hydrocarbons.

We do not offer any details of the elaborate reaction schemes that have been proposed but cite the following outline. The initial substances are actual 'gasoline' fumes and an array of derived substances resulting from incomplete combustion of gasoline. The derived substances include olefins, aldehydes, ketones, and possibly others. Readers desiring to pursue details are referred, for example, to Cadle and Allen (1970). The initial substances then react further under two primary influences: oxidation by atomic oxygen and activation by sunlight. The action of sunlight may dissociate the complex molecules into free radicals which are in turn much more chemically active. The various intermediate products may also attach to NO_2. One end product that has been widely believed to result is peroxyacylnitrate, $CH_3(CO)OO\ NO_2$, commonly abbreviated PAN. This rather strange substance is an eye irritant and is at least a strong enough oxidizing agent to oxidize NO to NO_2. It is potentially one of the irritants of smog although it is not clear that it is also a condensation nucleus for smog droplets.

Very recently, the chloride ions arising from ocean spray are reported to react with amine and ammonia radicals to form chloramine, a substance actually used in tear gas. The key to much of the chemistry involved falls back to the formation of organic radicals by sunlight, and these in turn are chemically active. For a review of the complex and poorly understood role of hydrocarbons in pollution see Altshuller and Bufalini (1971).

10. Allergens and Allergies

The human body, and doubtless many animal types, is a complex organic system that contains highly intricate defensive mechanism against foreign substances. In general, the defenses are against foreign proteins although the defensive mechanism has been known to come into play against such a substance as aspirin.

The body is initially inert to any given foreign substance until it receives an actual exposure to it. Such a first exposure (depending on its strength and duration) does not ordinarily induce reactions, but it does *sensitize* the body to the substance. Thereafter, a subsequent and frequently a very minute exposure triggers the protection mechanisms, often out of proportion to the magnitude of the exposure. This phenomenon is called an *allergic reaction*. The mucous irritation, sneezing, or bronchial constriction are the result of the generation of histamine or related substances in excess amounts in the blood stream. They are the consequences of the reaction to the foreign substance – or allergies – rather than the primary irritation by the allergen proper. Aero-allergens are estimated to result in a greater impairment of health than all of the air pollutants being controlled (Eisenbud, 1970).

Plants and animals and even microorganisms generate a vast and diverse amount of material that finds its way into the atmosphere and that can potentially trigger reaction in human beings. There is virtually no limit to the number of offensive substances. Molds are probably the most poorly understood allergens and most resistant to anti-allergy drugs. Whether the notorious *ragweed allergy* is so common because of unusual human sensitivity to it or because the abundance of ragweed pollens in the atmosphere causes extensive exposure and sensitizes even relatively resistive individuals, the fact remains that there is a very great deal of ragweed and many people become sensitized to it. A drastic reduction in ragweed level would materially reduce the incidence of 'hay fever'.

In some persons allergic reactions occur to substances that are useful and beneficial in general. (There are individuals who are hypersensitive to egg whites, orange peels, wheat, cow's milk, etc.). Any considerations of control of allergens must therefore be confined to those having no known value. While ragweed is indeed a weed, its ecological function would need to be studied before widespread destruction should be undertaken. It would appear that studies of at least one of the most widespread and most commonly irritating allergens (ragweed pollen) could be undertaken with profit even if at great cost.

The allergenic material transmitted by air is invariably solid, particulate matter. As a consequence, air conditioning with its filtering and precipitation processes is effective in reducing the level of allergens, with noticeable benefit to 'hay fever' victims. This fact supports the 'threshold' picture of allergic reaction; with allergenic levels below threshold for a given individual, no reactions at all occur.

Air conditioner manufacturers, salesmen, and engineers have been pointing out for nearly 40 yr that human beings pass some 35 lbs of air through their lungs in 24 hr, almost ten times the amount of food and liquid they consume, yet they exercise far less control and spend less money per individual on pure air control than on food and water purity. Air conditioning in office areas and homes doubtless reduces the incidence of allergic reactions to airborne pollutants.

11. Radioactivity

A short comment on the subject of radioactivity is included here because it is both a natural and a man-made component of the atmosphere and because concerns have grown lest the man-made component grows excessively.

Radioactive substances exist in the atmosphere and have done so long before man-made A bombs or H bombs were exploded. Natural sources include radon and thoron gases which decay to ^{210}Pb, cosmic radiation, and radioactive elements produced by cosmic radiation such as ^3H, ^{40}K, ^{14}C, ^{22}Na, ^7Be, etc. The natural radioactive gas, radon, may constitute the major part of the natural, airborne component. Human beings are exposed to the radiations from radioactive substances and to cosmic rays, and particularly at higher altitudes to energetic radiations from astronomical sources. The average dose to the overall population of the United States is 90 to 100 m rem yr^{-1} (milli-roentgen-equivalent-man). The unit called a roentgen was defined many years ago as the radiation that would release 1 electrostatic unit of charge (hence about 3.33×10^{-10} C) from a cm^3 of air at NTP exposed to the radiation with sufficient electrostatic field applied to cause saturation collection of the charges but without multiplication. Normally, such a field is around 300 V cm^{-1}. The modification of the roentgen to the roentgen-equivalent-man is introduced to add the specification that (1) the air ionization must be by 200 keV X-rays, and (2) the radiation must be measured by absorption and ionization in human tissues and must be compared with that by 200 keV X-rays.

For comparison and reference some figures are cited. A patient commonly receives 5000 to 7000 rem (note: *not* m rem) in a full cancer treatment. The radiation is limited very closely to the critical area. A total body exposure of 400 rem is regarded as the level for 50% chance of fatality. For many years, a complete gastrointestinal X-ray examination exposed the patient to about 70 rem but modern techniques are lowering this exposure. A simple radiographic picture of a hand, for example, exposes the patient to 0.5 to 2 rem depending on the procedure, but this is limited to the hand and is not total body radiation. What with dental and other X-rays, the average person in the United States receives 55 m rem per year (note: m rem) for medical purposes, but some receive very much more. The 55 m rem per year is to be added to the natural exposure of 90 to 150 m rem yr^{-1} making 145 to 205 m rem as the seemingly unavoidable total.

The International Commission on Radiation Protection has set a figure of 170 m rem yr^{-1} as the maximum permissible exposure for the population at large. It is clearly difficult if not impossible to go below this level. For single individuals, as opposed to a statistical average of the population at large, the same commission sets 500 m rem or 0.5 rem yr^{-1} as the maximum and 5 rem yr^{-1} as the upper limit for workers in radiation areas (hence presumably subject to continual health monitoring.)

As in many problems involving human beings, the sensitivity to radiation is variable from individual to individual as are also the characteristics of the damaging effects. Whether it is proved or not, it is highly likely that the human population falls onto a Gaussian distribution curve in respect to its radiation sensitivity so that occasional individuals are hypersensitive and others remarkably resistant. Whether a threshold level exists for any individual is not known. The decisions of the International Commission on Radiation Protection are based on *linear* extrapolation obtained from exposures of mice and other mammals to doses of 100 rem and more. Linear extrapolation is in itself an assumption as to the relation between dose and effect. One concomitant of it is that there is no such thing as a 'safe' dose, but only that the risk vanishes when the exposure vanishes. While this is certainly the safest assumption, a more reasonable one is that radiation destroys cells and the body replaces them, much as in the case of a burn, and that genuine toxicity to average individuals does not set in below a certain minimum. The Gaussian distribution law then describes how many individuals have an onset level at progressively lower and lower exposures. Experimentally, the difficulty in establishing any views about toxicity of low level radiation exposures lies in the essential abnormality of high sensitivity, its rarity, and hence the increased possibility of other, non-radiative causes of symptoms. (For example, leukemia is relatively rare, and its exact causes are not certain. Severe exposure to radiation has been known to be followed by leukemia. If a patient exhibits leukemia and has a history of very low dosages of radiation, it is not adequately proved that a correlation exists or that the leukemia might not have had another origin.)

If an individual eats, breathes, or otherwise absorbs a radioactive substance into his body, he is actually implanting a source of radiation and as a consequence is exposed to radiation for a protracted period. The termination of the exposure is fixed by one of two factors, whichever is the shorter. The radioactive material has a decay time so that its activity falls off exponentially with time. The substance, as a chemical, has a biological elimination time from the body. The latter fact leads to the use of 'radioactive tracers' to track the bodily absorption and elimination of chemicals. (The late Professor Ernest O. Lawrence used to perform a lecture demonstration of holding his hand over a Geiger-Mueller counter and eating a sample of salt containing a trace of radioactive Na. The counter started rapid counting in a matter of minutes, disclosing transmission of the salt into the blood stream, and then decayed as the salt in the blood stream was replaced.)

Elements of the second group of the periodic table, particularly Ca, Sr, Ba, and Ra, are biologically taken into bone structure and have a long biological half-life, approaching that of the individual. Hence the intake of radioactive forms of these elements, notably ^{90}Sr and ^{226}Ra (normal radium), is especially dangerous.

There are several industrial hazards of radioactivity through air pollution. Machining of radioactive materials can lead to the inhalation of dust. Processing of radioactive gases can lead to radiation exposure directly or through inhalation. The mining and processing of U ore to metal is an example. Through gas or dust there may be a hazard from ^{222}Rn and through dust there is a hazard from ^{235}U and ^{238}U. The established AEC procedures are quite detailed and adequate for all such hazards.

For ingestion of natural uranium the critical organ is the kidney. The 'Body Burden' for an 'uncontrolled' person is about 5×10^{-4} μCi, which corresponds to about 7.5×10^{-4} g of natural uranium (NBS Handbook, 1963). Particles whose diameters are 10 μm or larger cannot be inhaled into the lower respiratory tract. Assuming the density of the ingested particles to be 19 g cm^{-3}, the number of particles of a given size which must be taken into the person's system to equal the allowable burden are 7.5×10^4 for 10 μm particles of U and 1.2×10^3 for 40 μm.

Nuclear reactors are a concern for air pollution. Reactor cooling fluids do not appear to become significantly activated during their exposure to the reactor radiations. By the 1980's, nuclear power plants may produce the majority of the electric power in the industrial nations. The operation of the plants and the fuel processing can by accident result in all of the fission isotopes being released to the atmosphere; however, such accidents are not discussed herein. Three of the radioactive gases produced in reactors are of possible concern: ^{85}Kr, ^{3}H, and ^{131}I. ^{85}Kr is a fission product with a half-life of 10.76 yr, tritium ^{3}H is produced by the capture of 2 neutrons by an H atom and has a half-life of 12.5 yr, and ^{131}I is a fission product, but only has a half-life of 8 days. Considering all factors the main isotope of concern is ^{85}Kr. Eventually safety standards may result in the limitation of the amount of reactor power production in a given region; although much can be done to prevent the escape of ^{85}Kr into the atmosphere.

12. Heavy Metals

Many heavy metals or their compounds are particularly toxic to human beings, other animals, and plants. Not much is known about their distribution in nature and their role in air pollution. They are certainly industrial hazards. At least Pb and Hg are currently of concern.

Mercury. In certain compounds Hg is particularly poisonous, historically being employed for suicide and murder. The uncompounded Hg in the form of a liquid is no hazard. Large quantities can be swallowed without harm. Many mercury compounds have been employed for medicinal purposes for thousands of years. Soluble compounds like mercuric chloride ($HgCl_2$) are poisons whereas mercurous chloride (Hg_2Cl_2) is insoluble and is not toxic. Mercury can act through inhalation of the vapor. The vapor pressure at room temperature is already some 200 times the safe level. Mercury is commercially found in an ore called cinnabar in the form HgS. The extraction of the Hg by heating can put dangerous amounts of Hg vapor into the atmosphere locally.

Methyl mercury compounds are particularly hazardous. These are $(CH_3)_2Hg$ and CH_3Hg^+ which can be produced by bacteria from Hg or unharmful compounds of Hg (Goldwater, 1971).

The amount of Hg in the ocean is about 10^{11} kg. Hammond (1971) estimates that all of the Hg released to the environment by man is only 10^8 to 10^9 kg. Apparently tuna specimens 90 yr old have as much as those caught today. While it appears that man is not affecting the world-wide level of Hg, it is apparently easy to create local hazards. The natural Hg cycle seems to be completely unknown at this time. In general Hg is more of a biospheric and hydrospheric hazard than for atmospheric pollution.

Lead. Human beings have been extracting and using Pb for 4000 to 5000 yr. Natural sources include dust and volcanoes. Lead is easily vaporized during melting and thus is an industrial hazard. The cycle of Pb in the atmosphere, oceans, biosphere, etc., is poorly understood. It is estimated the most Pb in the atmosphere is in soluble form, either as a halide salt as emitted from motor vehicles or as Pb sulfate from the reaction of PbO_2 and SO_2.

It is interesting to note the amount of Pb per year found in the Greenland ice. For about 2000 yr before 1725 the Pb content was quite small. Beginning in 1725, there was a steady increase until about 1948, the yearly concentration in 1948 was 25 times that of 1725. Beginning in 1948 the Pb concentration steadily increased much more rapidly each year such that by 1970 it was about 15 times what it was in 1948. The increase in 1725 is attributed to industrialization and the increase in 1948 to the greatly enhanced use of Pb in gasoline.

Motor vehicles now use about 4×10^8 kg of Pb yr^{-1} which is emitted in small particles in the exhaust. The particulate matter emitted by motor vehicles is approximately 40% Pb. The size distribution shows that 85% of the particles are $<4\mu$. The maximum in the particle size distribution is at 0.25μ, with the 25 percentile points at 0.15μ and 0.40μ. In addition to the Pb from motor vehicles there is about 2×10^8 kg of Pb particulate from industrial smelting.

The natural background is calculated to be 1×10^{-9} μg cm^{-3} (Robinson and Robbins, 1971). Much more extensive monitoring data for Pb are required. Although the atmospheric content is highly variable and is very dependent on the meteorological conditions, certain trends in the concentrations are found as shown in Table XI.

TABLE XI

Pb concentrations

Type of region	μg cm^{-3}	Fraction in Pb
Urban	1×10^{-6}	1×10^{-2}
Nonurban	1×10^{-7}	2×10^{-3}
Remote	2×10^{-8}	1×10^{-3}

The mass fraction of the total aerosols in Pb also decreases as the distance from urban regions increases. The relative concentrations of the natural aerosols increase. In concentrated motor vehicle traffic, the amounts of Pb in very local regions may be 10 to 50 times higher than the urban average.

The concentration of Pb falls off rapidly with distance from concentrations of motor vehicles. Most of the Pb aerosol is near to the Earth's surface and is easily removed by gravity, rain, and snow. The average lifetime of the particles is a matter of days. Pb has been detected in the Greenland ice because of the extremely low background and because the prevailing winds blow from northeastern U.S.A. to Greenland in only a few days.

Most of the Pb aerosols deposited on the surface appear as fine black dust and are mostly soluble and readily washed away by the rain.

13. Particulates

Aerosols and particles play a very important role in determining the scattering properties of the atmosphere. They can produce a visible haze or smog. Atmospheric pollutants may be transported from one place to another as aerosols. Atmospheric aerosols may have been injected as particulates or they may have been emitted as gases and subsequently evolved into aerosols. The aerosol content of the atmosphere is found to vary drastically in space and time.

Most of the particulate matter in the atmosphere is in the troposphere. Values are sensitive to the location of local sources and meteorological conditions. Each region with an air quality problem must be monitored to determine the quantity and content of the particulate in that particular region. These tropospheric particles can be considered in several different size groups. Those larger than 10 μm come from mechanical processes, erosion, grinding, spraying, and meteors. The size group of 1 to 10 μm is most numerous and comes from mechanical and industrial dust, combustion ash, and meteors. The 0.1 to 1 μm particles are condensation products. Particles below 0.1 μm are called the Aitken nuclei. They are the condensation nuclei and are of terrestrial origin and only seem to exist in the troposphere and can be from motor vehicles.

The atmosphere contains water droplets and ice crystals. These depend on the cloud conditions but are maximized at 5 to 15 μm and have a distribution of 2 to 40 μm. They do not seem to be a part of the pollution problem. It is not even known if H_2O plays a crucial role in photochemical or other pollution problems. Water droplets and rain are very important in scavenging aerosols from the atmosphere.

There are industrial dust hazards, such as silicosis from stone cutting. Safety procedures for most of these are understood. Currently, however, little is known about the hazard from loose asbestos (Eisenbud, 1970). Common asbestos is chrysotile, hydrated magnesium silicate. $Mg_6(OH)_8Si_4O_{10}$, and is mined from serpentine rock. Loose asbestos is used extensively for insulation to fill voids, cement, and is sprayed on structural steel as a fire retardant material. The atmosphere is contaminated during application and when the building is demolished. It is known that inhalation of asbestos can result in cancer. The time between inhalation and cancer is 10 or more years. Much more must be learned about this hazard.

As shown in Table VII over 90% of the particulate matter in the atmosphere results from natural sources. Most of the particulates remain in the troposphere and 80% are within the first km above the ground. Thus, most aerosols are a local effects problem. The natural sources of primary aerosols are sea salt, dust storms, forest fires, volcanoes, and meteors. Except for the meteoric dust and magnitude 1 volcano these aerosols do not get into the stratosphere. Primary man-made particle emission results from combustion (about 70% and most of this from burning coal) and industrial operations. These man-made sources are only a few % of the natural sources and do not affect the stratosphere.

The natural gaseous sources of aerosols are mainly S compounds, N compounds, and hydrocarbons and are far more important on a global basis than are man-made sources. Forest and vegetation produce more aerosols from the terpenes emitted than does man from his gaseous emissions. Except in certain urban regions man cannot compete with natural sources of aerosols and can do little about the blue haze that persists over much of the continents.

Particles must get into the stratosphere to show global effects. The natural aerosol layer is at 20 to 27 km altitude. There are three main sources for this aerosol layer: ammonium sulfate, volcanic dust, and meteoric dust. Except for a few years after a magnitude 1 volcano the stratospheric dust consists mostly of the sulfates and meteoric dust, which are distinctly different from each other (Rosen, 1969). The sulfates are produced from natural and man-made sources of H_2S and SO_2; however, on a world-wide basis the natural sources are much larger than the man-made sources. The larger particles are from meteors. The vertical distribution shows that they are falling from above. The meteoric dust is estimated to be 5×10^8 to 5×10^{10} kg yr^{-1}. The smaller particles are $(NH_4)_2SO_4$ or $(NH_4)HSO_4$ and are of 0.1 to 2μ radii. They are volatile and their distribution with height shows that their source is in the troposphere. These volatile particles could not exist above about 50 km altitude. They are produced by oxidation of H_2S and SO_2 migrating upwards. Figure 9 shows the particle size distribution of the sulfates and meteoric particles in the stratosphere.

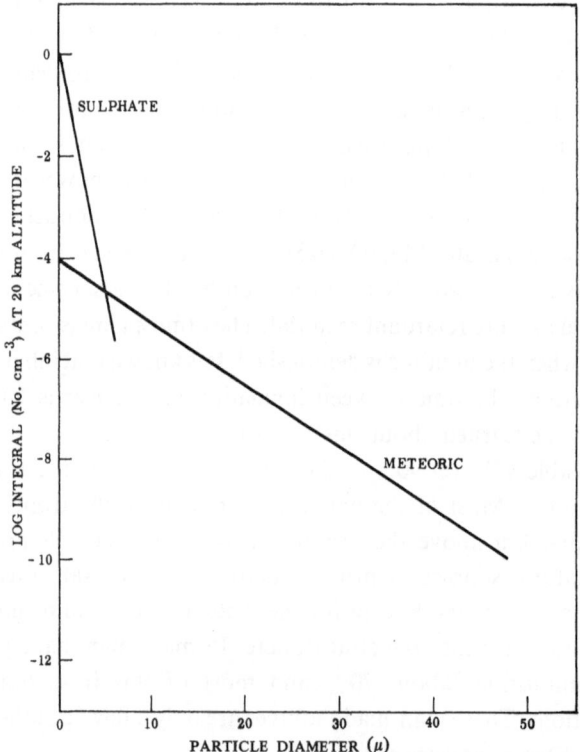

Fig. 9. Particle diameter distribution for sulfate and meteoric particles at the particle maximum in the stratosphere.

Figure 10 shows the mass fraction of sulfate and meteoric particles in the atmosphere as a function of altitude (Ory and Gilmore, 1971). The mass fraction for meteoric particles is essentially independent of altitude, while the sulphate particles show a maximum in the stratosphere.

Periodically volcanoes have injected large quantities of pollutants into the atmosphere. A large explosive type eruption can eject material high into the atmosphere,

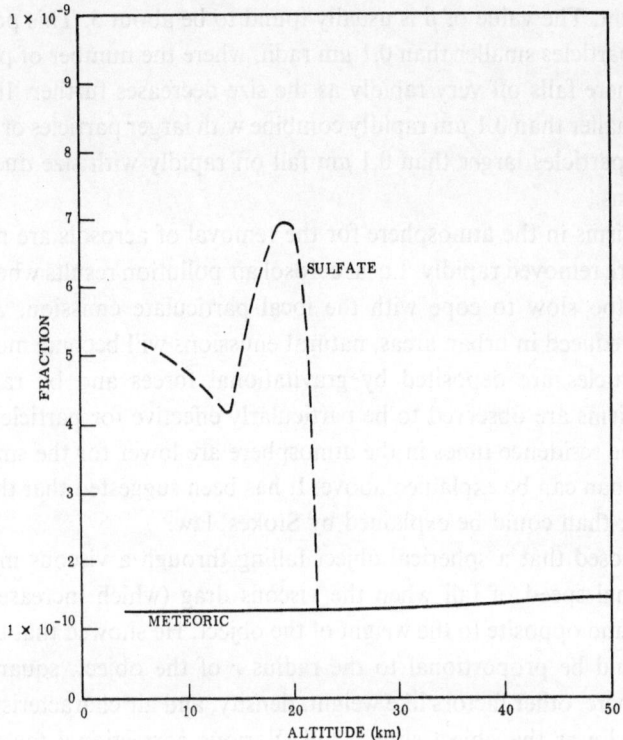

Fig. 10. Mass fraction of sulfate and meteoric particles as a function of altitude.

to at least as high as 50 km altitude. When ejecta is placed into the stratosphere, effects are observed world-wide.

It is estimated that Krakatau in 1883 ejected 10^{13} kg of dust into the stratosphere (Lamb, 1970). Optical phenomena as a result of this dust were observed for over 3 yr. Until 1963, the dust from volcanoes was studied from the dust deposit in ice and from visual accounts of the dust veil. After the Mt. Agung, Bali, eruption in March 17, 1963, dust samples in the stratosphere were collected and other scientific observations were made. Perhaps 10^{10} to 10^{11} kg of dust were injected into the stratosphere. Eruption particles of 0.5 to 0.8μ diam were found at ~20 km altitude. These were insoluble, dense, angular and were covered with water soluble material, part of which was sulfuric acid. For 80 to 150 days after the eruption, a concentration of 0.1 volcanic particles cm^{-3} was found at 20 km. It is estimated that the volcanic dust from Bali increased the stratospheric concentration by a factor of 2. It is estimated that 50 to 70% of volcanic dust is composed of SiO_2.

Although the distribution of aerosols is extremely variable, it has been found that in the lower troposphere the following power law can be used to approximately describe the size distribution

$$N = - Cr^{-\beta} \tag{48}$$

where N is the number of aerosol particles smaller than radius r, C is a constant and

β is the exponent. The value of β is usually found to be about 3. This power law does not apply for particles smaller than 0.1 μm radii, where the number of particles found in the atmosphere falls off very rapidly as the size decreases further. It appears that the particles smaller than 0.1 μm rapidly combine with larger particles or to form larger particles. The particles larger than 0.1 μm fall off rapidly with size due to enhanced loss mechanisms.

The mechanisms in the atmosphere for the removal of aerosols are rather efficient and particles are removed rapidly. Local aerosol air pollution results when the removal processes are too slow to cope with the local particulate emission. As man-made emissions are reduced in urban areas, natural emissions will become more important.

Aerosol particles are deposited by gravitational forces and by rain and snow. These mechanisms are observed to be particularly effective for particles >1 μm.

However, the residence times in the atmosphere are lower for the smaller particles (0.1 to 1 μm) than can be explained above. It has been suggested that the fall velocity is much higher than could be explained by Stokes' law.

Stokes proposed that a spherical object falling through a viscous medium should reach a terminal speed of fall when the viscous drag (which increases with speed) became equal and opposite to the weight of the object. He showed that terminal speed of fall v_f should be proportional to the radius r of the object, squared. Thus, the smaller the sphere, other factors like weight, density, and air characteristics remaining constant, the slower the object should fall. Various correctional factors have been proposed until the present 'best form' is believed to be

$$v_f = \frac{2}{9} g r^2 \left(\frac{e - e_a}{\eta} \right) \left(1 + \frac{B}{rp} \right).$$

The symbols are:

v_f = terminal speed of fall
r = radius of spherical particle
g = acceleration of gravity
e = density of particle
e_a = density of atmosphere
η = coefficient of viscosity of air
p = pressure of air
B = a constant for any given temperature. Its value, if p is in millibars and r is in cm is 0.0062 at $-55\,^{\circ}$C and 0.0084 at $+23\,^{\circ}$C. (Humphrey, 1940; Lamb, 1970).

When particles or droplets are as small as 1 μm, the value of v_f becomes measurable in terms of a kilometer in many weeks. In this case, the particles may well be said to 'float' since they move much faster under normal air currents than they fall according to Stokes' law.

Observationally the residence times are found to depend on altitude being $\sim$3 days at 1 km altitude and about 30 days in the upper troposphere. Much more data are required to improve our knowledge of aerosol loss mechanisms. See Robinson and Robbins (1971) for a detailed discussion of aerosol losses.

A very detailed analysis has been made by Lamb (1970) on the effects of volcanic activity on weather. He concludes that no 10 yr, 100 yr or ice age effects can be blamed on volcanic activities, and that many of the coldest winters in history occurred during periods of minima for volcanic dust in the stratosphere. Volcanic dust in the stratosphere does seem to reduce temperatures in the lower atmosphere for short periods.

There is still very much to be investigated about particulate matter in the atmosphere. Much detail is given by Cadle (1966).

References

Abelson, P.: 1971, 'The Environmental Crisis', *Trans. Am. Geophys. Union*, **52**, 124.
Altshuller, A. and Bufalini, J.: 1971, 'Photochemical Aspects of Air Pollution: A Review', *Environ. Sci. Tech.* **5**, 39.
American Chemical Society: 1969, 'Cleaning Our Environment, the Chemical Basis for Action', A Report by the Subcommittee on Environmental Improvement, Committee on Chemistry and Public Affairs, Washington, D.C.
Bolin, B.: 1970, 'The Carbon Cycle', *Sci. Am.* **223**, 124.
Bricard, J.: 1971, 'Formation of Aerosols in the Presence of SO_2 and NO_2', submitted to *J. Geophys. Res.*
Brosset, C. and Marsh, K. J.: 1970, 'Measurements of Sulphur Dioxide over the North Sea', *Atmospheric Environ.* **4**, 225.
Cadle, R. D.: 1966, *Particles in the Atmosphere and Space*, Reinhold Publishing Company, N.Y.
Cadle, R. D. and Allen, E. R.: 1970, 'Atmospheric Photochemistry', *Science* **167**, 243.
Charlson, R. J.: 1969, 'Atmospheric Visibility Related to Aerosol Mass Concentration', *Environ. Sci. Tech* **3**, 913.
Cloud, P. and Gibor, A.: 1970, 'The Oxygen Cycle', *Sci. Am.* **223**, 110.
Eisenbud, M.: 1970, 'Environmental Protection on the City of New York', *Science* **170**, 706.
Environmental Quality, Superintendent of Documents, USGPO, August 1970.
Gattinger, R. L.: 1971, 'Interpretation of Airglow in Terms of Excitation Mechanisms', in *The Radiating Atmosphere* (ed. by B. M. McCormac) D. Reidel Publishing Company, Dordrecht-Holland, p. 51.
Goldwater, L. J.: 1971, 'Mercury in the Environment', *Sci. Am.* **224**, 15.
Hammond, A. L.: 1971, 'Mercury in the Environment: Natural and Human Factors', *Science* **171**, 788.
Handbook of Geophysics and Space Environments: 1965 (ed. by S. L. Valley), McGraw-Hill Book Co., New York.
Humphrey, W. J.: 1940, *Physics of the Air*, McGraw-Hill Book Company, N.Y.
Hunt, B. G.: 1966, 'Photochemistry of Ozone in a Moist Atmosphere', *J. Geophys. Res.* **71**, 1385.
Hunten, D.: 1971, 'Airglow Introduction and Review', in *The Radiating Atmosphere* (ed. by B. M. McCormac), D. Reidel Publishing Company, Dordrecht-Holland, p. 3.
Jaffe, L. S.: 1970, 'The Global Balance of Carbon Monoxide', in *Global Effects of Environmental Pollution* (ed. by S. F. Singer), D. Reidel Publishing Company, Dordrecht-Holland, p. 34.
Johnson, F. S.: 1970, 'The Oxygen and Carbon Dioxide Balance in the Atmosphere', in *Global Effects Environmental Pollution* (ed. by S. F. Singer), D. Reidel Publishing Company, Dordrecht-Holland, p. 4.
Lamb, H. H.: 1970, 'Volcanic Dust in the Atmosphere: with a Chronology and Assessment of Its Meteorological Significance', *Roy. Soc. London Phil. Trans.* **266**, 425.
Latimer, W. M. and Hildebrand, J. H.: 1929, *Reference Book of Inorganic Chemistry*, Macmillan Company, N.Y.
McClatchey, R., Fenn, R. W., Selby, J., Garing, J. and Volz, F.: 1970, 'Optical Properties of the Atmosphere', AFCRL-70-0527, Bedford, Mass., U.S.A.

Nat. Bur. Stand Handbook: 1963, *Maximum Permissible Body Burdens and Maximum Permissible Concentrations of Radioactive Nuclides in Air and in Water for Occupational Exposure*, **69**.

Newell, R.: 1971, 'The Global Circulation of Atmospheric Pollutants', *Sci. Am.* **224**, 32.

Ory, H. A. and Gilmore, F. R.: 1971, 'Possible Effects of Atmospheric Particles on Fireball Emissions', R-669-DASA, Rand, Santa Monica, Calif., U.S.A.

Robinson, E. and Robbins, R. C.: 1968, 'Sources, Abundance, and Fate of Gaseous Atmospheric Pollutants', Final Report, SRI Project 6755 for American Petroleum Institute.

Robinson, E. and Robbins, R. C.: 1969, 'Sources, Abundance, and Fate of Gaseous Atmospheric Pollutants', Supplement, SRI Project 6755 for American Petroleum Institute.

Robinson, E. and Robbins, R. C.: 1970a, 'Gaseous Nitrogen Compound Pollutants from Urban and Natural Sources', *APCA J.* **20**, 303.

Robinson, E. and Robbins, R. C.: 1970b, 'Gaseous Sulfur Pollutants from Urban and Natural Sources', *APCA J.* **20**, 233.

Robinson, E. and Robbins. R. C.: 1971, 'Emissions, Concentrations, and Fate of Particulate Atmospheric Pollutants', *Am. Petroleum Inst.*, Pub. No. 4076, Washington, D.C.

Rosen, J. M.: 1969, 'Stratospheric Dust and its Relationship to the Meteoric Influx', *Space Sci. Rev.* **9**, 58.

Sax, N. I.: 1963, *Dangerous Properties of Industrial Materials*, Reinhold Publishing Company, N.Y.

SCEP: 1970, 'Man's Impact on the Global Environment', a report of the Study of Critical Environmental Problems, sponsored by MIT (MIT Press, Cambridge, Mass.).

USDHEW: 1966, Air Quality Data from National Air Sampling Networks and Contributing State and Local Networks, 1964–65.

Bibliography

Cadle, R. D.: 1966, *Particles in the Atmosphere and Space*, Reinhold, New York.

Cooke, L. M., Ed.: 1969, Cleaning our Environment: the Chemical Basis for Action, A Report by the Subcommittee on Environmental Improvement, Committee on Chemistry and Public Affairs. American Chemical Society, Washington, D.C.

Junge, C. E.: 1963, *Air Chemistry and Radioactivity*, Academic Press, New York.

Leighton, P. A.: 1961, *Photochemistry of Air Pollution*, Academic Press, New York.

Magill, P. L., Holden, F. R. and Ackley, C., Eds.: 1956, *Air Pollution Handbook*, McGraw-Hill, New York.

Singer, S. F., Ed.: 1970, *Global Effects of Environmental Pollution*, D. Reidel Publishing Co., Dordrecht-Holland.

Stern, A. C., Ed.: *Air Pollution*, 2nd Edition, Academic Press, New York, 3 vols.

Valley, S. L., Ed.: 1965, *Handbook of Geophysics and Space Environment*, McGraw-Hill, New York.

Webb, W. L.: 1966, *Structure of the Stratosphere and Mesosphere*, Academic Press, New York.

White, H. J.: 1967, *Industrial Electrostatic Precipitation*, Addison-Wesley, Reading, Mass., U.S.A.

Wilson, C. L., Ed.: 1970, *Man's Impact on the Global Environment: Report of the Study of Critical Environmental Problems (SCEP)*, The MIT Press, Cambridge, Mass.

AIR POLLUTION METEOROLOGY

R. W. SHAW and R. E. MUNN

Atmospheric Environment Service, Toronto, Canada

1. Introduction

An air pollution ecosystem is shown schematically in Figure 1. The size of the eco-system may be as small as a village or as large as the world. In all cases, the atmosphere is an integral component, transporting substances from sources to receptors and, on most occasions, diluting the pollution to acceptable levels. Dispersion rates are so

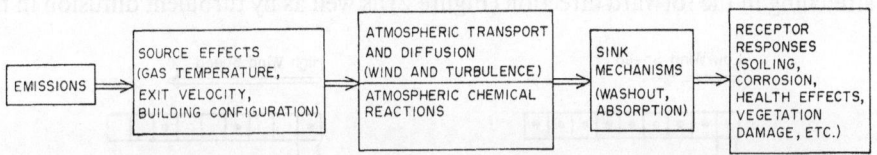

Fig. 1.   A flow chart showing the progress of air pollutants from sources to the receptors.

variable from place to place and from time to time, however, that the meteorological part of the ecosystem requires special study. Sometimes the natural ventilation provided by the atmosphere is very limited, and air quality deteriorates greatly. On other occasions the pollution mixes thoroughly through a very large volume of air.

A partial list of meteorological applications includes the following:

a) Design of chimney heights,
b) planning of industrial and regional land-use,
c) prediction of day-to-day air quality,
d) the selection of sites for urban and regional monitoring of pollution,
e) the prediction of receptor pollution uptake rates,
f) the interpretation of urban and regional trends in air quality (over decades),
g) the examination of relations between emission standards and air quality criteria.

A substantial body of theory exists on the transport and diffusion of pollution by a turbulent wind during so-called 'ideal' conditions, e.g., over uniform surfaces during steady-state flows. This theory is summarized briefly in Section 3. The models must be applied to real situations with great care, however, because the surface of the Earth is very irregular, and furthermore, the weather sometimes changes even from hour to hour. In general, reliable predictions of dilution rates cannot be obtained by simply applying formulas obtained from engineering handbooks. This is not because the formulas are wrong but because they cannot be safely extrapolated when the experimental situation has not been exactly replicated. In the irregular topography of the

McCormac (ed.), Introduction to the Scientific Study of Atmospheric Pollution, 53–96. All Rights Reserved.
Copyright © 1971 by D. Reidel Publishing Company, Dordrecht-Holland.

Rocky Mountains, for example, every chimney-height estimate is essentially a new research problem.

The development of air management strategies to combat pollution is interdisciplinary, but the meteorologist plays a key role in the search for optimum solutions. Some of the atmospheric models and methods are described in this Chapter, but no attempt is made to provide a 'cookbook' of practical formulas.

2. Classical Models

A. INTRODUCTION

A pollutant released into a calm atmosphere diffuses outward from the source because of the turbulent motion of the air. When a wind is blowing, the pollution is diluted by stretching in the forward direction (Figure 2) as well as by turbulent diffusion in the

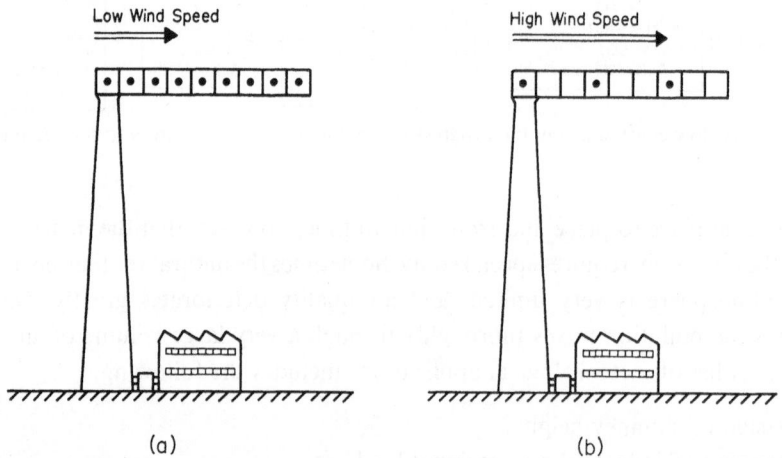

(a)   (b)

Fig. 2.   The higher the wind speed, the greater the volume of air that passes over the top of the chimney per unit time, and the greater the dilution of the effluent (Gifford, 1959).

crosswind and vertical directions. Thus the dispersion rates are governed by the strength of the wind and the nature of the turbulent field. In this section the two main approaches to the problem of turbulent diffusion, i.e., the gradient and the statistical theories, will be briefly outlined. A more detailed treatment is given by Pasquill (1962).

B. TYPES OF SOURCES

The concentration pattern depends partially upon the nature of the source. Examples of the most common types are listed below:

Type of source	Examples
Continuous point	plume from a factory stack
Continuous line	busy highway, crop-dusting aircraft
Continuous area	city
Volume	puff, cloud

It should be noted that the city is considered as an area source on the meso-scale, but it approximates a point source on a regional or global scale.

C. BASIC ASSUMPTIONS

(1) Uniform underlying surface, i.e., a homogeneous wind field in any horizontal plane;

(2) Steady-state conditions, i.e., no change with time in the strength or direction of the wind, or in the intensity of the turbulence;

(3) A passive pollutant, i.e., one that does not decay or interact within the atmosphere and that is not lost at the underlying surface by deposition or absorption.

By convention, an orthogonal coordinate system is used. The z-axis is vertical, with the surface of the Earth at $z=0$. The x-axis is in the direction of the mean wind, with the source at $x=0$, $y=0$, and $z=h$.

D. THE GRADIENT THEORY OF DIFFUSION

The earliest approach to diffusion was made by Fick (1855), a physiologist. He borrowed a concept from earlier studies in the conduction of heat and electricity: that the diffusive flux of a substance is proportional to the gradient of its concentration. From this argument the rate of change of concentration $\bar{q}$ of a diffusing substance at any point is given by,

$$\frac{\partial \bar{q}}{\partial t} = -\bar{u}\frac{\partial \bar{q}}{\partial x} + \frac{\partial}{\partial x}\left(K_x \frac{\partial \bar{q}}{\partial x}\right) + \frac{\partial}{\partial y}\left(K_y \frac{\partial \bar{q}}{\partial y}\right) + \frac{\partial}{\partial z}\left(K_z \frac{\partial \bar{q}}{\partial z}\right). \tag{1}$$

The first term on the right-hand side of Equation (1) is the advection in the x direction due to the mean wind $\bar{u}$. The quantities inside the brackets are the diffusive fluxes in the x, y and z directions. They are proportional to the gradients of concentration; the K's inside the brackets are called the diffusivities. Except in very light winds, the second term on the right side of Equation (1) can be neglected.

It has been observed that when the value of the molecular diffusivity of air is used for K_x, K_y, and K_z, Equation (1) underestimates the rate of atmospheric diffusion by several orders of magnitude. The reason is that pollutants are diffused not only by molecular agitation, but to a much greater extent by the turbulent eddies in the lower atmosphere.

The problem of choosing values of diffusivities and solving Equation (1) under varying boundary conditions has been the subject of much work (for summaries, see Sutton (1953) and Pasquill (1962)). The simplest case is Fickian diffusion, which assumes that $K_x = K_y = K_z = K$ (a constant), i.e., that the turbulence is isotropic and stationary in time. As a simple illustration, consider Roberts' (1923) solution of Equation (1) for diffusion from a continuous point source at ground level. As described by Sutton (1953), the boundary conditions are

(1) $\bar{q} \to 0$ as x, y, $z \to \infty$

(2) $K_z \, \partial \bar{q}/\partial z \to 0$ as $z \to 0$, $x > 0$ (the ground is impervious to the pollutant)

(3) $\bar{q} \to \infty$ as $x \to 0$, $y \to 0$, $z \to 0$.

$$(4) \quad \int\limits_{-\infty}^{\infty}\int\int \bar{u}\bar{q}(x,y,z,)\, dx\, dy\, dz = Q$$

where Q is the total amount of pollutant emitted per unit time. Condition (4) states that matter is neither lost (deposited) nor created in the plume. The solution,

$$\bar{q}(x,y,z) \approx \frac{Q}{4\pi x (K_y K_z)^{\frac{1}{2}}} \exp\left[\frac{-\bar{u}}{4x}\left(\frac{y^2}{K_y} + \frac{z^2}{K_z}\right)\right] \qquad (2)$$

is shown for the x and y directions in Figure 3. Note that there is an inverse decrease of concentration with distance x along the axis of the plume ($y=z=0$) while at a given x, the distributions are Gaussian in the y and z directions.

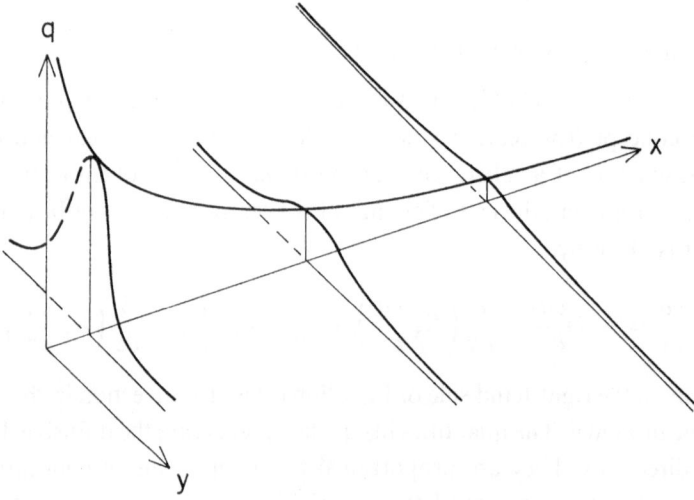

Fig. 3.   The downwind (x) and crosswind (y) distributions of concentration q in a plume from a continuous point source.

In actual fact, the assumption of Fickian diffusion does not apply in the lowest layers of the atmosphere because the eddy diffusivity changes with height. Sutton (1947) observed smoke and gas over level terrain from a ground-based continuous point source, and he found that the observed decrease of concentrations with downwind distance was much less than that predicted by Equation (2), indicating that K was indeed not constant.

The great popularity of the gradient, or K theory as it is sometimes called, stems from the fact that it affords the opportunity to obtain explicit solutions for $\bar{q}$ by assuming height-dependent forms of K and $\bar{u}$. Much elegant work has been done in two dimensions, i.e., for a continuous line source. However, the difficulty of formulating universal and realistic forms of K (particularly of K_z) has caused meteorologists to turn to another approach, that of using the statistics of the turbulence itself to estimate the diffusion.

E. THE STATISTICAL THEORY OF DIFFUSION FROM A CONTINUOUS POINT SOURCE

In a homogeneous field of turbulence, the diffusive spread of tagged particles from a continuous point source is a random process. Thus the distributions of particles in planes aligned orthogonally to the mean wind may be assumed to be approximately Gaussian, by the Central Limit Theorem of statistics. This model, with the assumptions listed in Section 2c, leads formally to the equation,

$$\frac{\bar{u}\bar{q}}{Q} = \frac{1}{2\pi s_y s_z} \exp\left(-\frac{y^2}{2s_y^2}\right)\left[\exp\left(\frac{-(z-h)^2}{2s_z^2}\right) + \exp\left(\frac{-(z+h)^2}{2s_z^2}\right)\right] \quad (3)$$

where s_y and s_z, the standard deviations of the particle distributions in the y and z directions, respectively, are functions of downwind distance x, as well as of the spectrum of turbulence. The double exponential in the final bracket embodies the assumption that there is perfect reflection at the surface of the Earth. The only remaining problem, and not an inconsiderable one, is to preduct the behavior of s_y and s_z.

The coordinate reference frame for Equation (3) is fixed, with origin at the point source, and is called an Eulerian coordinate system. Most of the theoretical studies of diffusion, however, originated with Taylor's theorem, in the so-called Lagrangian system in which tagged particles are followed for given lengths of time. The motion of a single particle or fluid element cannot be predicted, but some inferences can be made about the probability distributions of an ensemble of such particles.

Taylor (1921), the pioneer in the statistical approach, considered the displacement $x(t)$ of a single tagged particle as it is carried away from the origin by a turbulent wind $\bar{u}(t)$. Some scientists are familiar with this approach as the 'Random Walk' describing Brownian motion. If $x^2(t)$ is the mean-square displacement after a fixed time t, then,

$$\frac{dx^2(t)}{dt} = 2x(t)\frac{dx(t)}{dt} = 2x(t)u'(t)$$

Now

$$x(t) = \int_0^t u'(t+\xi)\,d\xi$$

Therefore,

$$\frac{\overline{dx^2(t)}}{dt} = 2\int_0^t \overline{u'(t)u'(t+\xi)}\,d\xi$$

$$= 2\overline{u'^2}\int_0^t R_L(\xi)\,d\xi \quad (4)$$

where $R_L(\xi)$ is the Lagrangian correlation coefficient between the velocity of the single particle at time t and that at time $t+\xi$, i.e.,

$$R_L(\xi) = \frac{\overline{u'(t)\, u'(t + \xi)}}{\overline{u'^2}}$$

and $\overline{u'^2}$ is the mean-square turbulent velocity. Thus,

$$\overline{x^2(t)} = \overline{2u'^2} \int_0^t \int_0^t R_L(\xi)\, \mathrm{d}\xi\, \mathrm{d}t. \tag{5}$$

The quantity $[\overline{x^2(t)}]^{1/2}$ is the one-dimensional standard deviation of the probability distribution of the particle positions after some given time t. The analysis is extended to three dimensions by introducing separate expressions for $[\overline{y^2(t)}]^{1/2}$ and $[\overline{z^2(t)}]^{1/2}$.

Without determining the general form of $R_L(\xi)$, two limiting cases of Equation (5) are immediately obvious. For close-in diffusion (t small), $R_L(\xi)$ is close to unity because turbulent velocities close together in time are highly correlated. For this case:

$$\overline{x^2(t)} \simeq \overline{u'^2} t^2. \tag{6}$$

For long periods, $R_L(\xi)$ approaches zero because turbulent velocities separated by long time intervals are practically uncorrelated. In this case,

$$\overline{u'^2} \lim_{t \to \infty} \int_0^t R_L(t_1)\, \mathrm{d}t_1 = K_1$$

and

$$\overline{x^2} \simeq 2K_1 t. \tag{7}$$

These results, although interesting, cannot be applied in any practical way because the ranges of validity of the small t and of the large t regimes cannot be stated *a priori*. There also remains the problem of relating these Lagrangian statistics to the Eulerian diffusion process (particles released serially from a point source and samples at fixed downwind distances). Because of the assumed steady-state field of turbulence, the quantity $\overline{u'^2}$ is the same in both systems, but the other variable in Equation (5), $R_L(\xi)$, needs to be evaluated. The statistics $\overline{x^2}$, $\overline{y^2}$ and $\overline{z^2}$ may be equated to s_x^2, s_y^2 and s_z^2, respectively (see Equation (3)) by assuming equivalence of space and time averages.

Ideally, the Lagrangian coefficient $R_L(\xi)$ can be determined by measuring, as a function of time, the position of an identifiable object such as a constant-height balloon which faithfully follows the motion of the air. In fact, what is more easily measured is the Eulerian correlation coefficient $R_E(t)$ from the wind velocity fluctuations at a fixed point in the flow. Gifford (1955) made a direct comparison of Lagrangian properties (from the motion of constant-height balloons) and Eulerian properties (from anemometers mounted on a tower) and found that $R_L(\xi)$ was displaced from $R_E(t)$ toward longer time lags. Subsequently, Hay and Pasquill (1959) made the simple assumption that $R_L(\xi) = R_E(t)$ where $\xi = \beta t$. The quantity β depends slightly on the intensity of turbulence but seems to have a value of about 4. There is thus a method of connecting the diffusive spread of a plume to the turbulence measured at a point, assuming that the turbulent field does not change in time or space. These assumptions are rarely realized in the atmosphere but the model has been found to give useful predictions under 'ideal' conditions.

Of particular importance too is the sampling time, which often is not sufficient to encompass the broad range of eddy sizes in the atmosphere. As downwind distance increases and/or as the sampling time increases, the diffusion is governed more and more by the large-scale eddies. This is illustrated qualitatively in Figure 4. The shape

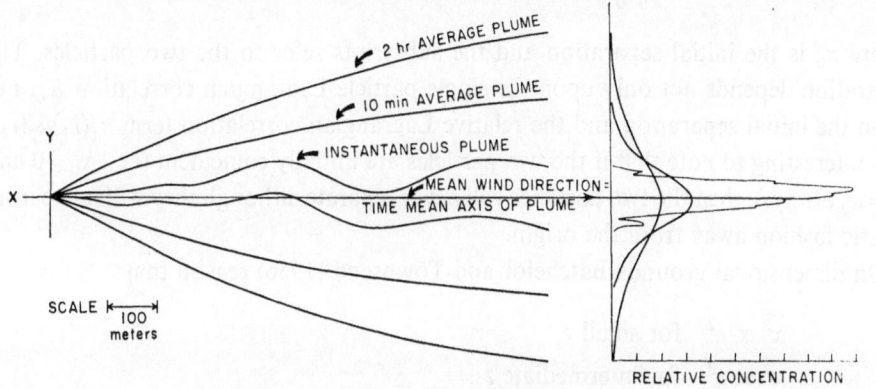

Fig. 4. The approximate outlines of a smoke plume observed instantaneously and averaged over periods of 10 min and 2 h. The diagram on the right shows the corresponding cross-plume distributions (U.S. AEC 1968).

of the instantaneous plume shows the effect of the small-scale turbulence fluctuations. As the time of averaging increases, the mean plume width increases. Pasquill (1962) has considered the implications of this, and his book is recommended for further reading.

Finally, it should be noted that Equation (3), as well as Equations (4) through (7), are based on the assumption of homogeneity. Under ideal conditions, this is realized in the cross-wind (y) direction. In the vertical, however, the wind increases with height above a solid surface, and the structure of the turbulence changes: large vertical fluctuations are suppressed near the ground. As a practical system, therefore, Equation (3) applies only to an *elevated* point source, and there has been new interest in recent years, e.g., (Gifford, 1962), in the K-theory (Section 2D) which can incorporate vertical gradients. The latter method predicts well the diffusive spread in the vertical direction over uniform surfaces, but still contains uncertainties in estimating the cross-wind diffusion.

F. THE STATISTICAL THEORY OF DIFFUSION APPLIED TO AN EXPANDING PUFF

Batchelor (1949, 1950, 1952) has considered the relative motion of two tagged particles in a Lagrangian field of turbulent motion. By studying the statistics of the separation of many such pairs, the distribution of pollutants in an expanding puff or cloud can be obtained. The mean square separation $\overline{x^2}$ is found from an argument similar to that for the one-particle case in the previous sub-section and is

$$\overline{x^2} = x_0^2 + \overline{2u'^2} \int_0^t \int_0^t R_L(t_2 - t_1) \, dt_1 \, dt_2$$

$$- 2 \int_0^t \int_0^t \overline{u_1'(t_1)u_2'(t_2)} \, dt_1 \, dt_2$$

where x_0 is the initial separation and the subscripts refer to the two particles. The separation depends not only upon the single particle Lagrangian correlation R_L, but upon the initial separation and the relative Lagrangian correlation term $u_1'(t_1)u_2'(t_2)$. It is interesting to note that if the two particles are initially coincident (i.e., $x_0 = 0$ and $u_1' = u_2'$ always), then the two particles will never separate although they will move in an erratic fashion away from the origin.

On dimensional grounds Batchelor and Townsend (1956) reason that

$$\overline{x^2} \propto t^2 \quad \text{for small } t$$
$$\overline{x^2} \propto t^3 \quad \text{for intermediate } t$$
$$\overline{x^2} \propto t \quad \text{for large } t. \tag{8}$$

In contrast to the one-particle dispersion described by Equation (5) through (7), the expanding puff contains an intermediate t^3-regime.

G. INTEGRATED PUFF MODEL OF A PLUME

The diffusion equations described above are appropriate for steady-state conditions over homogeneous surfaces. However, mesoscale features such as cities and lakes cause changes in the wind and the turbulence along the length of a plume. Plume dispersion under these conditions can be represented more realistically by the 'integrated puff' model in which the plume consists of a continuous train of discrete puffs which are carried downwind in a stepwise fashion according to the local wind (Figure 5). While they are travelling downwind, the puffs expand according to the local dispersion conditions, also in a stepwise fashion. Roberts et al. (1969) constructed a model of the dispersion of pollutants over Chicago by superimposing the pollution patterns from many sources (both commercial and residential), each of which was described by the 'integrated puff' model. They were able to model the three-dimensional variation of wind and atmospheric diffusion.

The concentration of pollutants within the puff may be assumed as a first approximation to be Gaussian in three dimensions, i.e.,

$$\bar{q} = \frac{Q}{(2\pi s^2)^{\frac{3}{2}}} \exp\left[\frac{-r^2}{2s^2}\right] \tag{9}$$

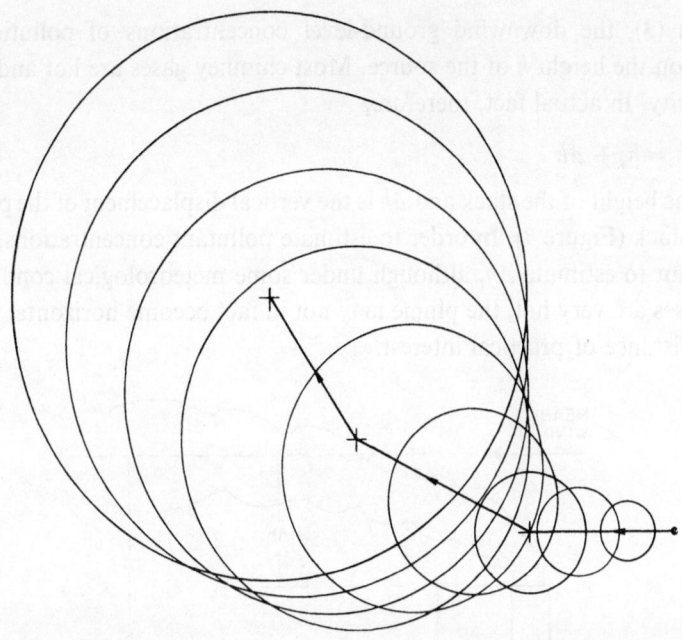

Fig. 5.   Plan view of several stages of a puff travelling downwind and expanding by diffusion. A
plume can be represented by a continuous train of such puffs. (Roberts *et al.*, 1969).

where the puff diffuses isotropically, the standard deviation is s, and r is the distance
from the center of the puff. However, in many cases the transport by the wind is
much greater than the diffusion in the downwind direction and the puff may be
replaced by an expanding disk (Figure 6). The disk travels with the wind and the
concentration across the disk is given by a two-dimensional Gaussian model. Of
course, the expanding puff rather than the expanding disk model would apply in
light wind conditions.

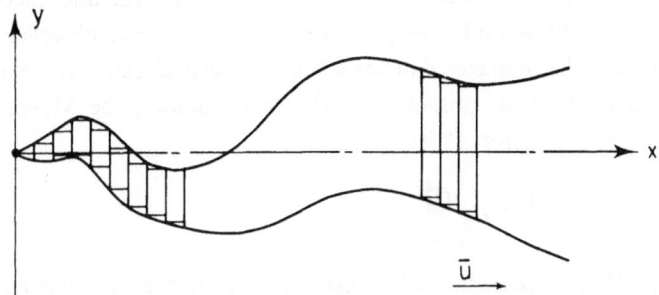

Fig. 6.   Similar to Figure 5, but for an expanding disc travelling downwind. This representation of a
plume shows the effect of meandering. (U.S. AEC, 1968)

H. BUOYANT PLUMES

In Equation (3), the downwind ground-level concentrations of pollution depend *inter alia* upon the height h of the source. Most chimney gases are hot and they have an exit velocity. In actual fact, therefore,

$$h = h_s + \Delta h$$

where h_s is the height of the stack and Δh is the vertical displacement of the plume upon leaving the stack (Figure 7). In order to estimate pollutant concentrations, therefore, it is important to estimate Δh, although under some meteorological conditions, and when the gases are very hot, the plume may not in fact become horizontal within any downwind distance of practical interest.

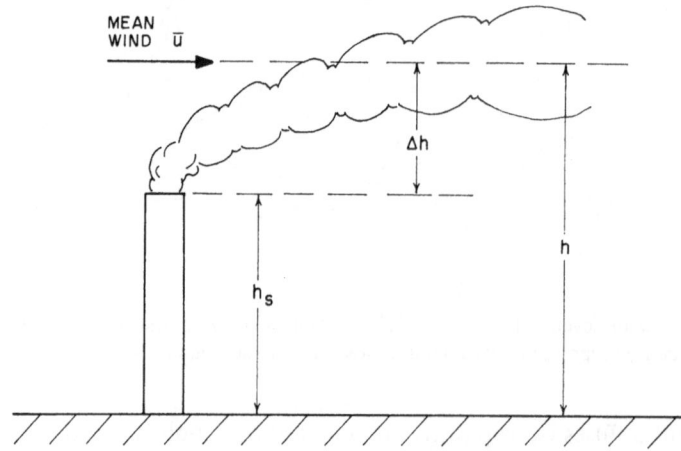

Fig. 7.   The rise of a plume above the top of a chimney.

The plume rise Δh depends upon the buoyancy (positive or negative) and the vertical momentum of the effluent, i.e., the exit velocity upon leaving the stack. Twenty or thirty plume-rise formulas have been derived, none of which has found universal application. A more detailed discussion of this problem is given in Section 6 and the reader is referred to summaries by Briggs (1969), and Carson and Moses (1969).

The formulas may be divided into two classes: empirical and theoretical. The empirical formulas rely upon least-square regressions between observed plume rises and factors such as the diameter d of the stack, the vertical exit velocity w_s, the mean wind speed $\bar{u}$ and the heat emission rate Q_H. For instance, the Moses and Carson (1968) all-data regression for Δh is:

$$\Delta h = C_1 w_s \frac{d}{\bar{u}} + C_2 \frac{Q_h^{\frac{1}{2}}}{\bar{u}} \tag{10}$$

where C_1 and C_2 are experimental 'constants', which are dependent upon atmospheric stability. The first term in Equation (10) accounts for the effect of vertical momentum; the second for the effect of buoyancy.

Because empirical formulas are applicable only for conditions similar to those under which they were derived, various theoretical approaches to the problem have been made by Batchelor (1954), Scorer (1959), Csanady (1961, 1965) and Briggs (1965), for example. These approaches have tried to model the turbulent entrainment of ambient air into a hot plume. Usually, the entrained air is less buoyant than the plume air and the rise is reduced (Figure 8).

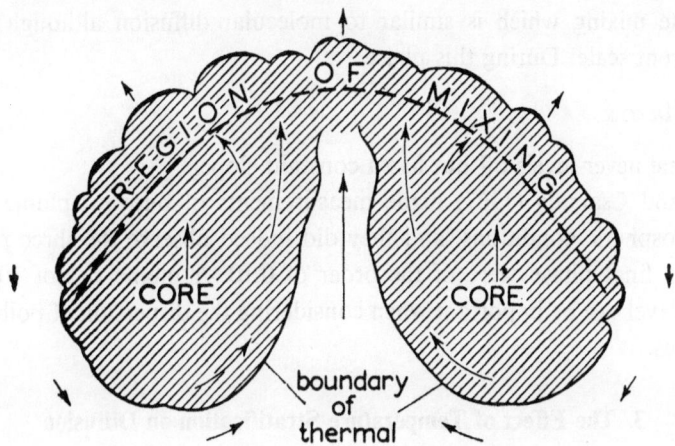

Fig. 8.   A buoyant plume acts as a mixing thermal or 'bubble' which mixes (entrains) with the environment. (Reprinted with permission from R. S. Scorer, *Natural Aerodynamics*, 1958, Pergamon Press, Ltd.)

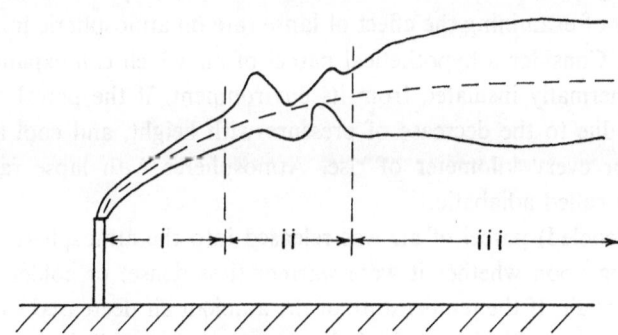

Fig. 9.   Three stages of plume rise according to Slawson and Csanady (1967). See text for details. Note that even in stage (iii) the plume does not completely level off.

Slawson and Csanady (1967) have divided plume rise into three phases (Figure 9). In each phase, they have postulated a different type of eddy diffusion:

(i) Initial phase: the turbulence generated by the vertical momentum of the plume itself dominates the mixing and the nature of the environmental turbulence is relatively unimportant. During this phase

$$\Delta h \propto x^{\frac{2}{3}}.$$

This result agrees with an earlier analysis of vertical plumes by Morton *et al.* (1956).

(ii) Intermediate phase: Intermediate size eddies in the environment, of a scale about the diameter of the plume, dominate the mixing. During this phase, the mixing is pronounced and there is a strong tendency for the plume to level off asymptotically, i.e.,

$$\Delta h \propto \log x.$$

(iii) Final stage: The energy-containing eddies, of a scale much smaller than the plume, dominate the mixing which is similar to molecular diffusion although of a completely different scale. During this phase

$$\Delta h \propto x$$

i.e., the plume never levels off but has a constant slope.

Slawson and Csanady (1967) made measurements of buoyant plume rise under various atmospheric conditions, and they did indeed observe the three phases. The slopes in the final phase were of the order of 0.10. A plume rise of 100 m/km of horizontal travel can be important when considering the dispersion of pollutants over long distances.

3. The Effect of Temperature Stratification on Diffusion

A. GENERAL CONSIDERATIONS

The decrease of temperature with height $(-\partial T/\partial z)$ is called the lapse rate, a quantity that is frequently used as an index of the dispersive capacity of the atmosphere.

A useful way of examining the effect of lapse rate on atmospheric mixing is by the parcel method. Consider a hypothetical parcel of air which can expand or contract but which is thermally insulated from its environment. If the parcel were lifted, it would expand due to the decrease of pressure with height, and cool at the rate of about 10°C for every kilometer of rise. Atmospheres with lapse rates of about 10°C km^{-1} are called adiabatic.

If a heated (cooled) parcel of air was released into the atmosphere, it would rise (sink) depending upon whether it were warmer (less dense) or colder (more dense) than the ambient air. If the temperature of the ambient air decreased with height at a rate which is greater than the adiabatic lapse rate (a super-adiabatic condition), the parcel would remain less dense than its environment and would continue to rise. The relationship between the temperature of the environment and the parcel is shown in Figure 10. Similarly, a descending parcel would warm by adiabatic compression at approximately 10°C km^{-1}, and it would continue to descend if the ambient lapse rate were superadiabatic. Thus, an atmosphere with a superadiabatic lapse rate is said to be unstable because vertical motion or 'convective overturning' is encouraged. In the unstable atmosphere, therefore, pollutants are easily mixed in the vertical.

By a similar argument, if the lapse rate is less than the adiabatic, the air is said to be stable because vertical motion is damped. Thus, in a stable atmosphere, pollutants are not easily mixed in the vertical. An isothermal condition is one in which there is

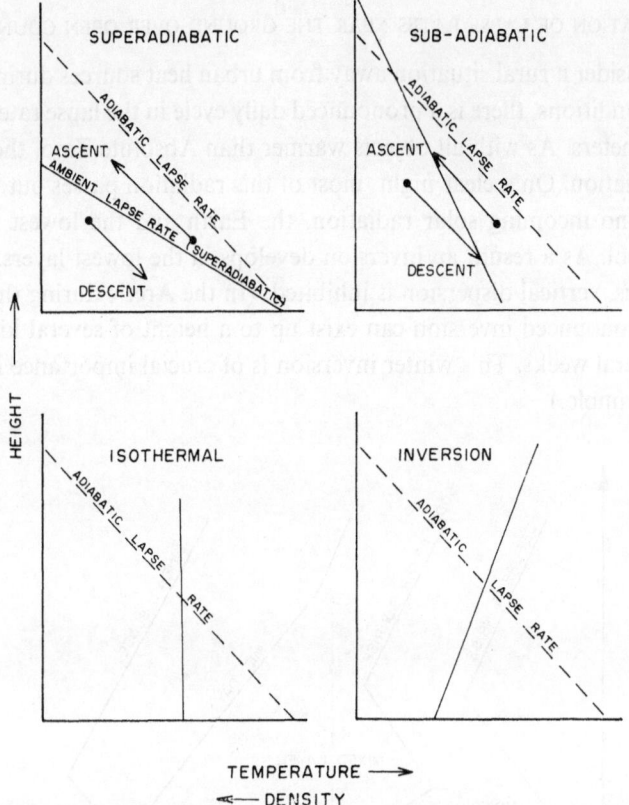

Fig. 10. The effect of ambient lapse rate (decrease of temperature with height) upon the buoyancy of a rising or descending parcel of air. Under superadiabatic conditions, vertical motion is encouraged. Under the other three conditions, vertical motion is hindered to varying degrees.

no change of temperature with height (moderate stability) while an inversion occurs when the temperature actually increases with height (strong stability and little vertical mixing).

Superadiabatic lapse rates occur mainly close to the ground, in the lowest few hundred meters of the atmosphere. They are associated with (a) sunny daytime conditions, (b) cold air advection over warm bodies of water, for example, over the Great Lakes in winter.

Neutral conditions occur (a) during cloudy, windy weather, (b) for brief periods just after sunrise and just before sunset, signalling the onset and termination of daytime thermal convection.

Inversions occur (a) at night near the ground during clear skies and light winds, (b) when there is warm air advection over cold bodies of water or melting snow surfaces, (c) after a summer shower when the ground is cooled by evaporation, (d) aloft at frontal surfaces separating cool from warm air masses, (e) in the middle troposphere (at heights of 1000 to 5000 m) in warm subtropical anticyclones (the Trade Wind inversion, for example), and (f) in the stratosphere.

Some of these cases will be described below.

B. DAILY VARIATION OF LAPSE RATES NEAR THE GROUND OVER OPEN COUNTRY

Let us first consider a rural situation away from urban heat sources during clear skies. Under these conditions, there is a pronounced daily cycle in the lapse rate in the lowest few hundred meters. As with all objects warmer than Absolute Zero, the Earth emits long-wave radiation. On a clear night, most of this radiation passes out to space and, since there is no incoming solar radiation, the Earth and the lowest layers of the atmosphere cool. As a result, an inversion develops in the lowest layers, as shown in Figure 11. Thus, vertical dispersion is inhibited. (In the Arctic during the polar night in winter, a pronounced inversion can exist up to a height of several kilometers and persist for several weeks. This winter inversion is of crucial importance in Fairbanks, Alaska, for example.)

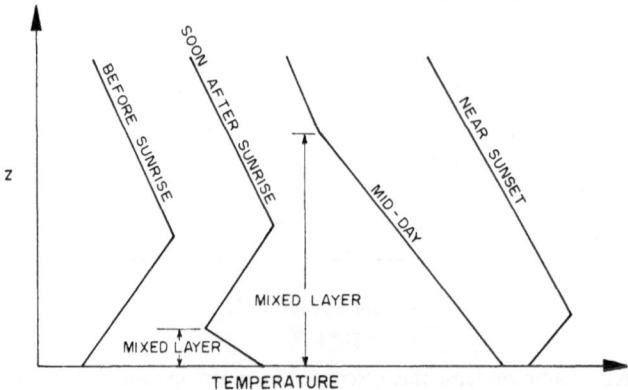

Fig. 11.   The distribution of temperature with height at several times throughout the day at a rural location under clear skies. The height of the mixed layer is a minimum before sunrise and a maximum in the afternoon.

After sunrise, the air near the ground is warmed and an adiabatic or super-adiabatic lapse rate develops. The night-time inversion is eroded from below by convective mixing. As this process continues, pollutants which may have collected *aloft* during the night are brought rapidly downward to Earth resulting in momentarily high concentrations. This process is known as a 'fumigation'. Eventually, (usually well before noon), the stable inversion is completely replaced by a well-mixed surface layer, which prevails until near sunset when the night-time inversion begins to develop again.

The top of the mixed layer is called the mixing height. It is an important quantity in many practical systems for forecasting day-to-day regional air quality. During fine weather in the country, the mixing height is zero at sunrise, increasing to a maximum value in the afternoon.

The effect of a cloud cover and/or of strong winds is to weaken the daily cycle. The cloud re-radiates long wave radiation back to Earth, with the result that the night-time cooling is lessened. During the day, on the other hand, the cloud reduces the incoming insolation and there is less warming of the surface layer. Strong winds tend to stir the air and reduce the vertical temperature gradients. During overcast windy conditions, in fact, the lapse rate is close to the adiabatic throughout both day and night.

C. URBAN EFFECTS

The rural temperature regime described above may be modified considerably by mesoscale features such as cities and lakes which influence the radiation balance. Studies in Louisville, Kentucky (De Marrais, 1961) and in Montreal, Canada (Munn *et al.*, 1963) show that the night-time inversion occurs less frequently in a city than in the surrounding countryside, as might be expected for the following reasons:

(1) Industry, transportation, and space heating act as heat sources.

(2) The large areas of concrete, brick, and asphalt absorb heat from the Sun during the day and release it gradually at night. The lowest layers of the air are slow to cool at night and the formation of the inversion is inhibited.

(3) The pall of smoke and carbon dioxide over a city acts as an infrared emitter/absorber, radiating both upward and downward. The upward radiation causes cooling at the top of the smoke layer; the downward radiation warms the city below. As illustrated in Figure 12 the effect of this is to retard the formation of the urban inversion.

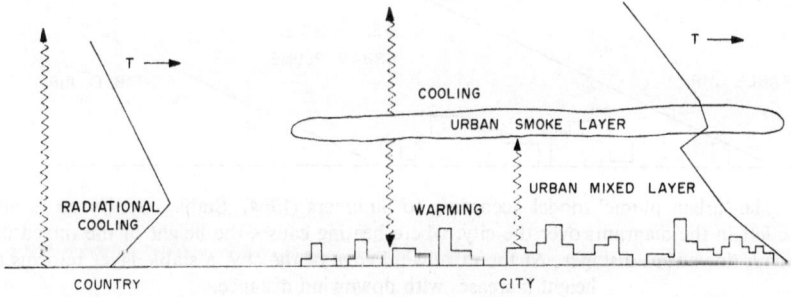

Fig. 12. The downward infrared radiation from the urban smoke layer warms the lower layers of air over the city and inhibits the formation of a nighttime inversion. The night-time mixed layer is, therefore, present in the city but not in the country during periods of clear skies and light regional winds.

(4) The resulting city 'heat island' induces a convective circulation when regional winds are light (Figure 13) with rising air over the city and subsiding air over the suburbs. A night-time inversion over the suburbs is enhanced by the subsidence; the inversion over the city is weakened by the rising air.

(5) Automobile traffic stirs the air near the ground and weakens the inversion.

The height of the urban mixed layer is an important quantity in many practical systems for forecasting urban air quality.

Summers (1964) has proposed a model for the air flow over a city at night as shown

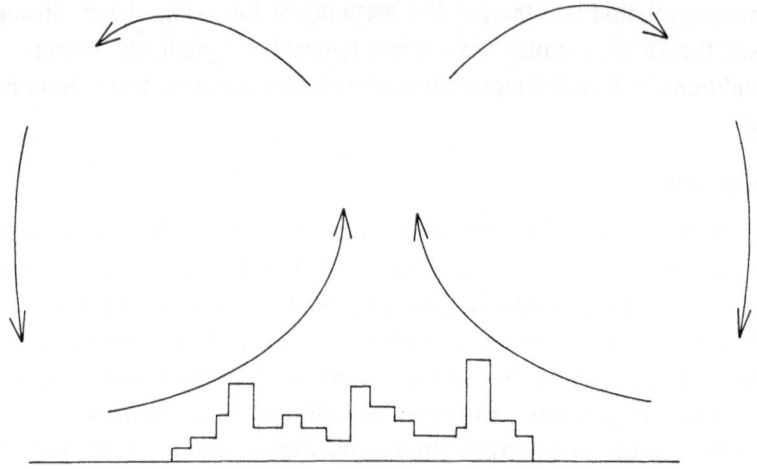

Fig. 13.   Under conditions of light regional winds and clear skies, the heating in the city causes
the air to rise. Descent takes place in the surrounding countryside.

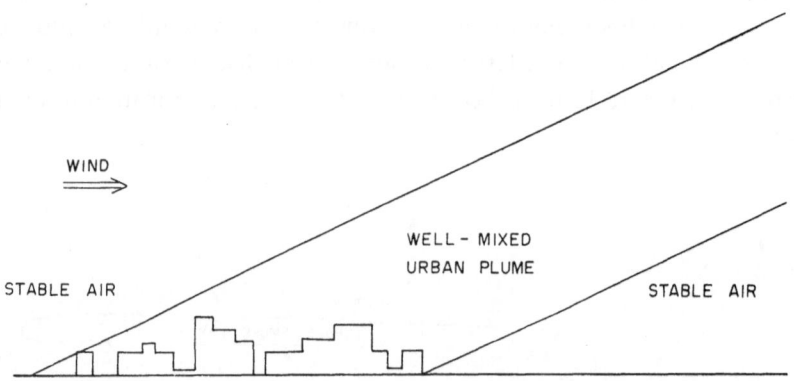

Fig. 14.   The 'urban plume' model according to Summers (1964). Stable country air is advected
(from the left in the diagram) over the city, where heating causes the height of the mixed layer to
increase with downwind distance. At the downwind edge of the city, a stable layer reforms and its
height increases with downwind distance.

in Figure 14. Stable country air crossing the city is modified from below, the thermal
and mechanical turbulence creating a surface well-mixed layer. The depth of this
mixed layer increases with distance downwind over the city. When the air passes
again from city to country, however, a low-level inversion reforms and deepens with
distance downwind. The region between the boundary layers is called the 'urban
plume'. Its existence has been verified experimentally by Clarke (1969) in Cincinnati
and by Oke and East (1971) in Montreal.

D. SHORELINE EFFECTS

Air passing from a large lake or the ocean to a land surface in spring and summer has
its temperature structure changed in the lowest layers, as shown in Figure 15. Frequently there is a temperature inversion over the lake due to the cool water. As the

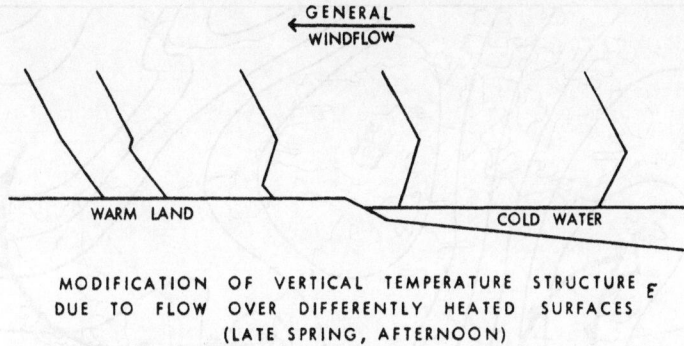

MODIFICATION OF VERTICAL TEMPERATURE STRUCTURE
DUE TO FLOW OVER DIFFERENTLY HEATED SURFACES
(LATE SPRING, AFTERNOON)

Fig. 15. The change in the lapse rate as air passes from cold water to warm land. The low level inversion over the water is destroyed by heating over the land (USDHEW, 1967).

air passes over the relatively warm land, heating from below creates an unstable surface layer, eroding the inversion from below. The unstable layer therefore deepens as the air proceeds inland. Pollutants which have collected aloft in the inversion layer may be brought to the surface by convective overturning. This 'fumigation' process can occur for extended periods of time, in contrast to the brief fumigation which occurs during the breakup of the night-time inversion.

In the autumn and winter, unfrozen bodies of water such as the Great Lakes are warmer than the adjoining land. The air over the water is frequently unstable (causing convective clouds and snow flurries). In a warm air mass with southerly winds, however, an inversion may form over the relatively cool downwind land surface, trapping pollutants emitted in the surface layer and resulting in high pollution levels.

E. LARGE-SCALE SYNOPTIC INFLUENCES

The surface weather map in Figure 16 shows lines of equal atmospheric pressure (called isobars). The heavier line with pips on it is the surface position of a front, a relatively sharp boundary between warm air to the south and cold air to the north.

As shown in Figure 17, the resultant of the pressure gradient force (PFG) and the Coriolis force (CF) due to the Earth's rotation causes the surface air to spiral counter-clockwise in the Northern Hemisphere around the center of a low pressure area and clockwise around the center of a high pressure area. Surface frictional drag (FF) deflects the motions slightly, and to satisfy continuity, air must rise in the cyclone and subside in the anticyclone. Rising air tends to cause an adiabatic lapse rate to be established with the result that vertical mixing is moderately encouraged in a low pressure area. In an anticyclone, however, a 'subsidence inversion' often forms at mid levels (2 to 3 km), inhibiting vertical mixing of pollutants. The warm subtropical anticyclones at latitudes 30°N and 30°S in the Atlantic and Pacific Oceans have the great potential for high pollution concentrations because they are quasi-stationary. One of the causes of the smog problem in Los Angeles, for example, is that the region is under the influence of the semi-permanent Pacific anticyclone. The mid-latitude cold continental anticyclones that are located between the migrating cyclones pose a

Fig. 16. A surface weather map with 'isobars' (light lines) which are loci of equal atmospheric pressure and 'fronts' (heavy lines) showing the boundary between warm and cold air masses. High pollution concentrations occurred in the stagnant high pressure area over the lower Great Lakes and Eastern seaboard (Copyright: Academic Press, New York).

less serious threat because they are rarely stationary. In addition, they undergo heating from below and are often unstable in their lowest layers. Nevertheless, these high pressure areas sometimes become stationary for several days, particularly in autumn in the southeastern United States, where they are reinforced by western extensions of the Bermuda-Azores anticyclone. Besides the subsidence inversions at 2 or 3 km, there is a permanent inversion at about 10 km where the tropopause acts as a lid on tropospheric pollution (except in the vicinity of jet streams).

Figure 18a shows a vertical cross section along line AA' (Figure 16), where the cold air is advancing. This 'cold front' has a slope of about 1 in 50. After the passage of such a front, there is often a startling improvement in the clarity of the air.

Figure 18b shows a vertical section along line BB' in Figure 16 through a warm front. This front has a shallower slope (about 1 in 200) than the cold one and because it moves over relatively cool ground, the air is stabilized. Particularly when the warm

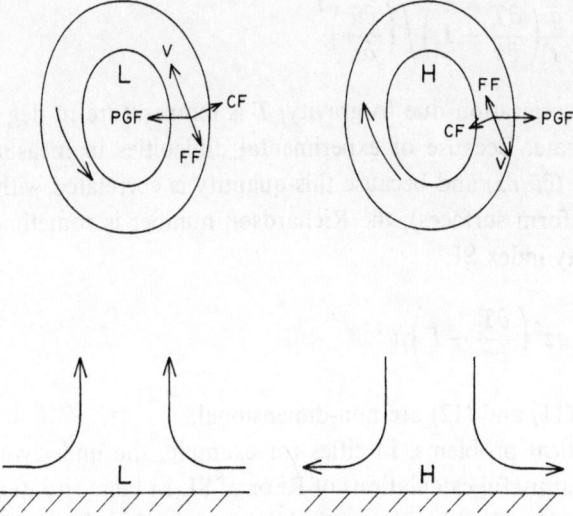

Fig. 17. The resultant of pressure (PGF), Coriolis (CF) and friction (FF) forces causes air to spiral inwards and rise in a cyclone (L), but to spiral outwards and descend in an anticyclone (H).

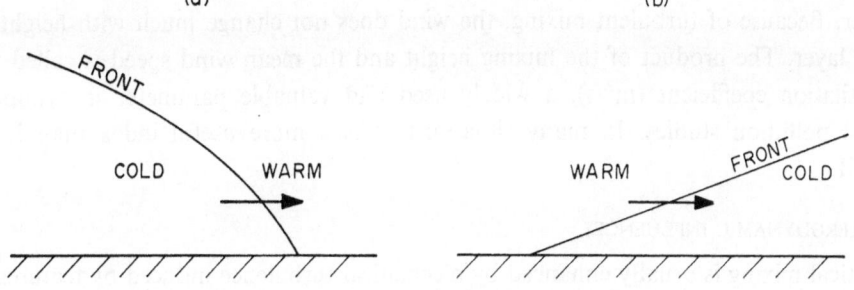

Fig. 18. Part (a) is a vertical section through a cold front (along line AA' in Figure 15). Part (b) is a section through a warm front (line BB' in Figure 15).

front is slowly moving or stationary, a serious pollution potential situation can develop. In hilly country the potential may remain high even after the warm front has passed, because pockets of cold air may linger in valleys. The overriding warm flow then acts as a lid to inhibit vertical diffusion.

4. The Effect of Wind on Transport and Diffusion

A. GENERAL CONSIDERATIONS

The lapse rate is only a qualitative index of turbulence. The structure of turbulence (and thus the rate of expansion of a plume) can vary widely, depending not only on the buoyancy regime but also on the strength of the wind and the roughness of the underlying surface. For precise experimental work over uniform surfaces, an appropriate stability parameter is the Richardson number Ri, which incorporates both buoyancy and wind effects.

$$\mathrm{Ri} = \frac{g}{T}\left(\frac{\partial T}{\partial z} + \Gamma\right) \Big/ \left(\frac{\partial \bar{u}}{\partial z}\right)^2 \tag{11}$$

where g is the acceleration due to gravity, T is temperature in deg K, and Γ is the adiabatic lapse rate. Because of experimental difficulties in measuring the vertical gradient of wind $(\partial \bar{u}/\partial z)$ and because this quantity is correlated with the wind speed $\bar{u}$ itself (over uniform surfaces), the Richardson number is sometimes replaced by a simplified stability index SI:

$$\mathrm{SI} = gz^2\left(\frac{\partial T}{\partial z} + \Gamma\right)\Big/\bar{u}^2. \tag{12}$$

Both Equations (11) and (12) are non-dimensional.

In many practical problems, in cities for example, the underlying surface is too irregular to permit useful calculations of Ri or of SI. In these situations, a simple box model is found to be valuable. The vertical dimension of the box is the mixing height, and for a sufficient downwind distance from the sources, a uniform vertical distribution of pollution can be assumed. The other meteorological element of interest is then the rate of horizontal ventilation, given by the mean wind speed through the mixed layer. Because of turbulent mixing, the wind does not change much with height in this layer. The product of the mixing height and the mean wind speed is called the ventilation coefficient (m^2/s), a widely used and valuable parameter in synoptic-scale pollution studies. In many situations, it is a more useful index than is Ri or SI.

B. AERODYNAMIC INFLUENCES

Vertical mixing is usually enhanced by mechanical turbulence induced by features in the terrain such as trees, buildings, and hills. However, the waffle-like configuration of buildings and streets in a city causes complex flow patterns near ground level that may produce local convergence of pollutants. Georgii et al. (1967), for example, studied the distribution of carbon monoxide in Frankfurt/Main and found that the concentration of CO was several times higher on the leeward side of the street than on the windward side. The patterns of CO concentrations and the associated air circulations are shown in Figure 19.

Figure 20 (Halitsky, 1962) displays schematically the air flow around a single building during strong winds. If a short chimney were to release pollutants below the surface of separation shown in Figure 20, high concentrations would be carried to the ground in puffs from time to time in the lee of the building. This kind of problem occurs also in the turbulent wakes of cliffs. Because the flow regimes are critically dependent on local geometry and on the angle of attack of the wind, the meteorologist cannot readily predict the resulting concentration patterns. Satisfactory engineering solutions are possible, however, by wind-tunnel simulation (provided that subsequently, a new tall structure is not erected in the vicinity, radically altering the flow patterns).

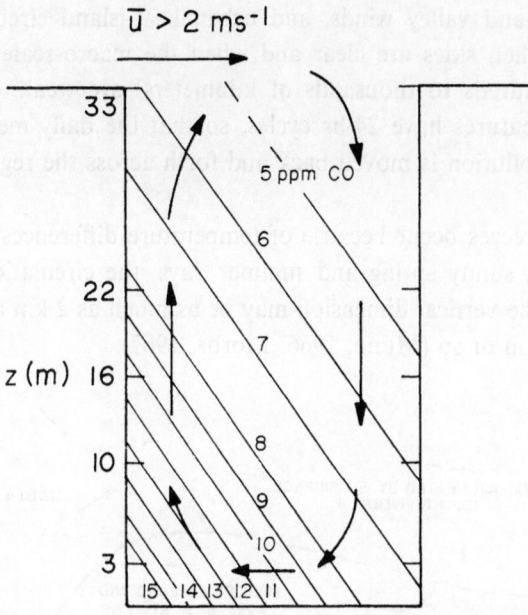

Fig. 19.   Vertical cross-section of a street and the tall buildings lining it, showing the circulation of air (arrows) and isopleths of concentrations of carbon monoxide. Note the high concentrations of CO at street level on the upwind side (Georgii *et al.*, 1967).

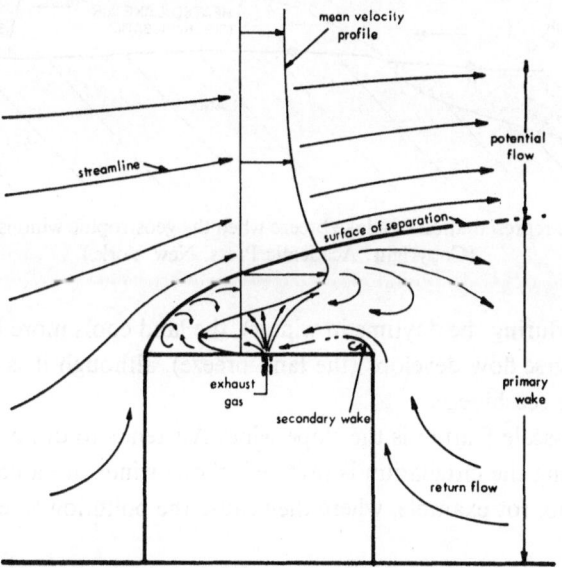

Fig. 20.   Typical flow pattern around a cube with one face normal to the wind (Halitsky, 1962). (Copyright: Academic Press, New York.)

C. MESO-SCALE WIND FLOWS

The dispersion of pollution is achieved not only by relatively small turbulent eddies (Section 2) but also by larger-scale organized circulations.

Meso-scale flows (horizontal dimensions of a few kilometers) include land-and-sea-breezes, slope and valley winds, and urban heat-island circulations. They are most noticeable when skies are clear and when the macro-scale flows (horizontal dimensions of hundreds to thousands of kilometers) are weak. Characteristically, these meso-scale features have 24-hr cycles, so that the daily mean *vector* wind is small. Thus, the pollution is moved back and forth across the region without much dilution.

Land and sea breezes occur because of temperature differences between the land and the water. On sunny spring and summer days, the circulation is as shown in Figure 21, where the vertical dimension may be as much as 2 km and the horizontal extent may be 20 km or so (Munn, 1966; Moroz, 1967).

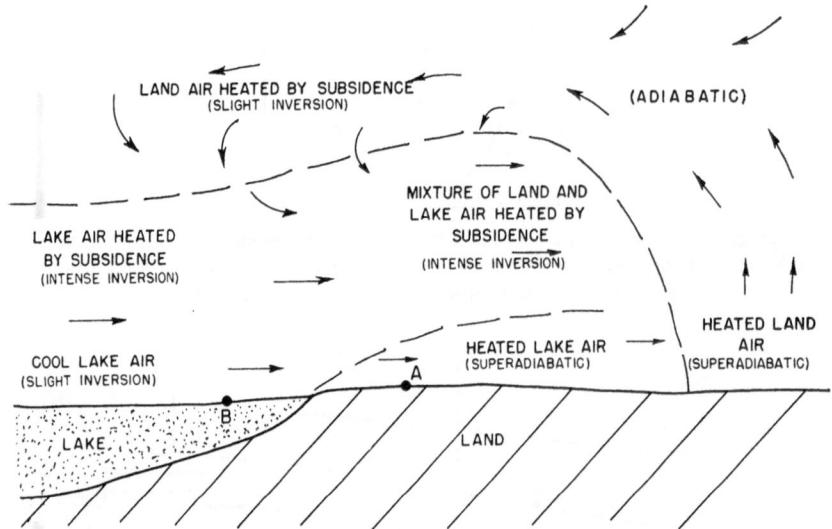

Fig. 21.   Schematic representation of a lake breeze when the geostrophic wind is light (Munn, 1966). (Copyright: Academic Press, New York.)

At night (and during the daytime in winter), the land cools more by radiation than the water. A reverse flow develops (the land breeze), although it is generally weaker than the daytime sea breeze.

Another meso-scale feature is the slope wind. Air tends to drain downhill at night but in the morning the circulation is reversed. Slope winds are a common feature in Denver, Colorado, for example, where they cause the pollution to ebb and flow over the city.

Valleys generate local flow patterns (in addition to slope winds) and create special problems for industrial siting. Figure 22 shows the wind regime and lapse rates in the Bushmans valley, oriented northeast–southwest in South Africa (Tyson, 1968). After sunrise, the western slopes are heated while the eastern slopes are still in the shade: a closed circulation results. These results cannot be generalized because of differences in local geometry, orientation, and the existence sometimes of several secondary valleys or canyons. A number of physical and numerical models have been proposed

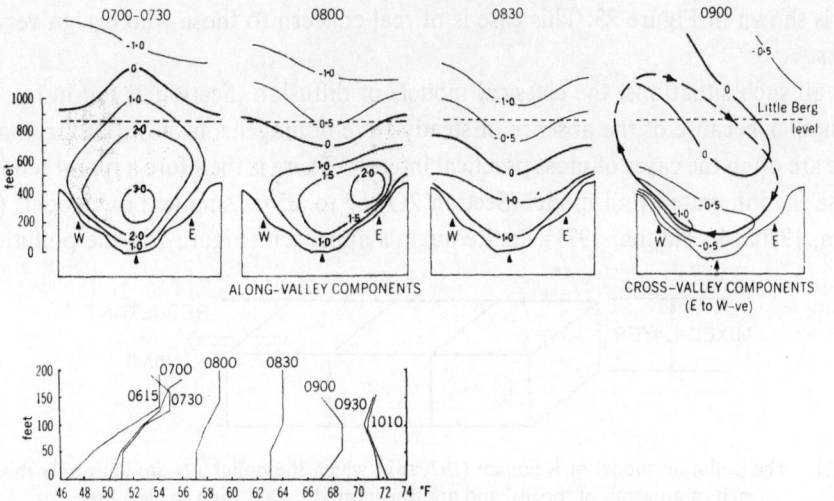

Fig. 22. (a) Vertical sections across a valley showing isopleths of wind speed (positive values are up-valley, or East-to-West, where applicable). (b) Distribution of temperature with height at mid-valley (Tyson, 1968).

for 'ideal' configurations, but they cannot be extrapolated readily to real situations. The advice of a meteorologist should be sought in each individual case.

Finally, reference has already been made in Section 3C (see Figure 13) to the urban circulations that develop when regional winds are light. The problem again is to find an ideal case: most cities are located on coasts, slopes, or in valleys so that the resulting flow patterns can be very complex. The position of the urban heat island in Toronto, for example, is affected by the lake breeze (Munn *et al.*, 1969).

During light winds the hot gases from even a single tall chimney can generate a closed circulation (Schmidt and Boer, 1963), causing the plume to loop down to ground level, sometimes several kilometers from the source. A schematic representa-

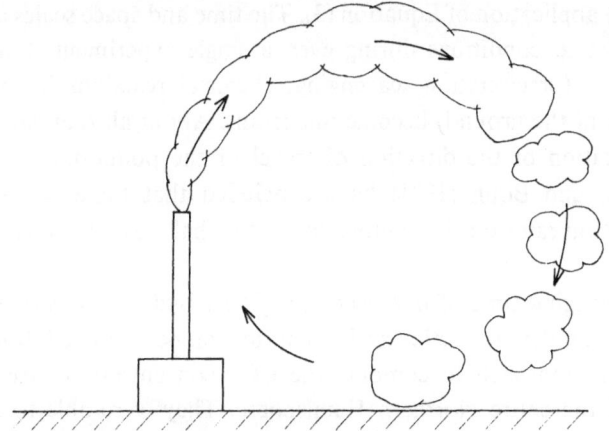

Fig. 23. During light winds, hot gases in a plume can create a circulation which can cause pollution to loop down to ground level some distance from the source.

tion is shown in Figure 23. This case is of real concern to those who design very tall stacks.

In all such situations, the classical models of diffusion (Section 2) fail in one way or another because of the absence of steady-state homogeneous flow. Unfortunately, these are often the cases of most practical interest. There is therefore a recent tendency to use the integrated puff model (Section 2G) or to adopt simple 'box' models (Reiquam, 1970a, b; Hanna, 1971). In Reiquam's approach (Figure 24), the pollution is

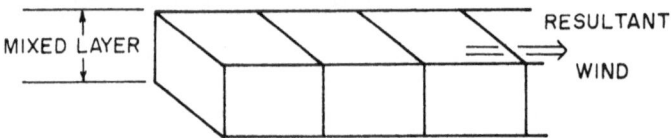

Fig. 24.   The pollution model of Reiquam (1970a) in which the pollutants are uniformly mixed in each of an array of 'boxes' and are transported by the wind in each box.

assumed to be uniformly mixed in each of an array of boxes. Net transport can be computed if information is available on the wind in each box, averaged over each of consecutive time periods. Predictions from Reiquam's model have compared favorably with observations in the Willamette Valley of western Oregon and in a small Norwegian valley.

D. MACRO-SCALE WIND FLOWS

The classical diffusion models can sometimes be applied to downwind distances of more than 100 km over uniform surfaces (open countryside or large bodies of water). Peterson (1968), for example, undertook aircraft monitoring of the Brookhaven argon-41 plume over the Atlantic to a distance of 300 km on a day when winds were steady from the west. He found that the cross-wind diffusion nomograms of Pasquill (1962) could be extrapolated to these distances. Over land, however, the irregular terrain and the diurnal rise and fall in the thickness of the surface mixed layer frequently prevent application of Equation (3). The time and space scales are too large to assume steady-state conditions during even a single experiment. Furthermore, the sink mechanisms (precipitation scavenging, chemical reactions in the atmosphere, and adsorption at the ground) become important. About all that can be obtained is a general indication of the direction of travel of the pollution, using a trajectory analysis. Munn and Bolin (1971) have concluded that there is no possibility of estimating dilution rates on the continental and global scales from classical diffusion theory.

Using an averaging time of one year, the global budgets of particular pollutants can be estimated, if the production and sink terms can be evaluated. Bolin and Bischof (1970) used this approach to compute the CO_2 concentrations averaged over the troposphere of the entire Northern Hemisphere. They were able to determine that the CO_2 concentrations are rising at the rate of 0.7 ± 0.1 ppm/yr. Although the method can in principle be used with any pollutant, there is frequently great difficulty in estimating one or more terms in the budget, particularly when natural sources such as the sea are as important as man-made sources.

For regional and continental evaluations, advection in and out of the area complicates the analysis, and additional research is required in most cases. A possible methodology is contained in the study by Reiquam (1970b) of the sulfur transport in western Europe during a simulated 10-day period of light southwesterly winds. Only man-made sources were considered and no account was taken of removal mechanisms. The resulting patterns (derived from the model shown in Figure 24) are displayed in Figure 25. This diagram cannot yet be verified and because sink

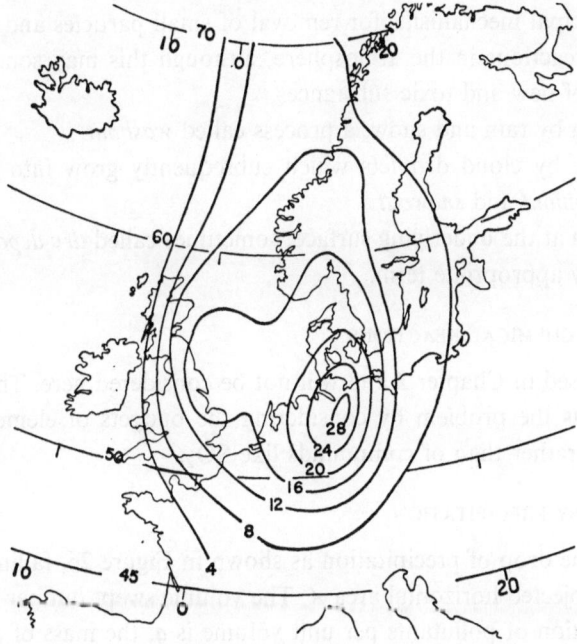

Fig. 25. Mean concentration of sulphur ($\mu g\,m^{-3}$) during a simulated air pollution episode in Western Europe (Reiquam, 1970b). (Copyright 1970 by the American Association for the Advancement of Science.)

mechanisms were ignored, the gradients may not be realistic. However, the possibility exists that regional patterns of pollution can be determined when the wind field is not too complex, when the mixing height is well defined, and when the source and sink strengths can be evaluated.

5. Pollution Sinks

A. GENERAL CONSIDERATIONS

The diffusion models generally assume that the pollution remains in the atmosphere, with perfect reflection at the underlying surface. This is usually a satisfactory assumption for travel distances of a few hundred meters in the absence of fog or precipitation.

Heavy particles such as fly ash are deposited, of course, by the action of gravity. This loss may be estimated by adding an empirical correction term to Equation (3). In any event, heavy-particle deposition is no longer an important practical consideration in air pollution control: industrial precipitators are capable of removing large particles before the effluent leaves the chimney. The atmosphere too is relatively efficient at cleansing itself of extraneous materials, and only during conditions of poor horizontal and vertical ventilation can a serious deterioration of air quality take place. On the global scale, for example, the rise in pollution concentrations over decades is only a fraction of what would be expected from a consideration of increased production alone.

The four principal mechanisms for removal of small particles and gases are,

(1) chemical reactions in the atmosphere, although this may sometimes result in the production of new and toxic substances,

(2) scavenging by rain and snow, a process called *washout*,

(3) scavenging by cloud droplets which subsequently grow into precipitation, a process called *rainout* and *snowout*,

(4) adsorption at the underlying surface, sometimes called *dry deposition*, although this is not a very appropriate term.

B. ATMOSPHERIC CHEMICAL REACTIONS

These are discussed in Chapter 2 and will not be considered here. The meteorologist sometimes avoids the problem by considering the budgets of elemental substances such as sulphur rather than of compounds like SO_2.

C. SCAVENGING BY PRECIPITATION

Consider first one drop of precipitation as shown in Figure 26, falling with speed w, and having a projected horizontal area A. The volume swept out per unit time is wA. If the concentration of pollutants per unit volume is q, the mass of pollutants intercepted per unit time is qwA. However, the mass of pollutants collected per unit time may be less or greater than that in the swept volume, due to turbulent escape out of or entrainment into the volume. By introducing a collection efficiency E, the mass scavenging rate formally becomes $EqwA$.

The collection efficiency E depends in a complicated way upon the sizes and surface properties (such as wetness) of the collector and of the collected particles. The collection efficiency of raindrops is discussed by Mason (1957), McDonald (1963), and Berg (1963). Very small ($< 1\mu$ diameter) pollutant particles tend to move out of the way of the scavenger; however, electrical attraction offsets this effect (Byutner and Gesina, 1963).

Relatively little is known about the collection efficiency of snow. Slowly falling snowflakes should be efficient collectors of small particles (Georgii and Weber, 1964; Perkins and Engelmann, 1966) but Facy (1962) suggests that they may in fact shed particles. In summary, the evidence is conflicting.

The scavenged volume depends upon the sizes of the precipitation drops. For rain,

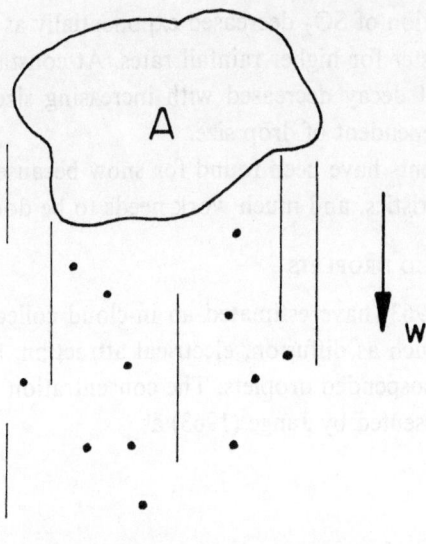

Fig. 26. A scavenging particle with a projected horizontal area A which is falling with vertical speed w through a polluted region.

the best known size distribution is that given by Marshall and Palmer (1948), i.e., $dn/dr = -\lambda N_0 \exp(-\lambda r)$, where dn is the number of raindrops with radii between r and $r + dr$, N_0 is a constant, and λ depends upon rainfall rate. The size distribution of snowflakes is not as well known as that for rain. Snow can take many crystalline forms (Nakaya, 1954), each of which has its own cross sectional area and fall speed characteristics.

The fall speed w of the precipitation particle depends upon its size and shape. Values have been determined experimentally for rain by Gunn and Kinzer (1949) and for snow by Nakaya and Terada (1934).

The total mass of scavenged pollutants is found by integrating over all N collector particles, i.e.,

$$\frac{dq}{dt} = \int_0^N E \, qwA dn.$$

The solution is an experimental decay of concentration with time

$$q = q_0 \exp(-S_A t) \tag{13}$$

where q_0 is the original concentration and S_A is the washout coefficient equal to $\int_0^N EwA \, dn$.

Some workers have attempted to evaluate the washout coefficient using a combination of theoretical and measured values for E, w, and A. Notable are the washout coefficients for rain by Chamberlain (1953) and May (1958). Beilke and Georgii (1968) allowed uniform-sized rain droplets to fall through a chamber filled with sulphur dioxide. The concentration of SO_2 decreased exponentially as predicted by Equation (13). The decay was faster for higher rainfall rates. At constant rainfall rates below 20 mm h^{-1}, the rate of decay decreased with increasing size of the drops. Above 20 mm h^{-1}, it was independent of drop size.

Few washout coefficients have been found for snow because of the lack of knowledge about its characteristics, and much work needs to be done still.

D. SCAVENGING BY CLOUD DROPLETS

Byutner and Gesina (1963) have estimated an in-cloud collection efficiency of only 0.1, and other effects such as diffusion, electrical attraction, and solubility in water become important for suspended droplets. The concentration K (mg l^{-1}) of in-cloud droplets has been represented by Junge (1963) as

$$K = \frac{\varepsilon q}{L}$$

where q is the concentration of pollutants in the air (g m^{-3}), L the liquid water content of the cloud (g m^{-3}), and ε is the rainout coefficient ($0 < \varepsilon < 1$). Beilke and Georgii (1968) estimated ε to be between 0.01 and 0.10 for SO_2, a range lower than that of 0.1 to 0.8 calculated by Junge (1963) for Aitkin nuclei (solid particles of radii 10^{-3} to 10^{-1} μ). Beilke and Georgii also made calculations of the rainout and washout of SO_2 and sulphates. The rainout (in-cloud scavenging) of SO_2 was small compared to the washout below cloud base because the assumed concentrations of SO_2 (based on observations in industrial areas) were much higher below a height of 1 km than they were in the clouds. In contrast, most of the capture of sulphates occurred within the cloud, rather than below it. Andersson (1969) analyzed the concentration of potassium, calcium, chlorine and sulphur in rainwater near the town of Uppsala, Sweden, and found that the maximum concentrations were in the town where the air was the most polluted.

Precipitation chemistry networks have been in existence in Europe for more than 20 yr. There are very few North American observations, however, and we can only speculate on the magnitudes of sulphur losses from large plumes by precipitation scavenging.

E. DEPLETION OF POLLUTION AT THE SURFACE OF THE EARTH

The vertical flux F of pollution to or from a homogeneous surface during steady-state conditions is given by,

$$F = - K \frac{\Delta q}{\Delta z} \tag{14}$$

where K is the atmospheric diffusivity (cm^2/sc) and $\Delta q/\Delta z$ is the vertical gradient of pollution. Sometimes Equation (14) is rewritten in terms of a transfer velocity V (cm s^{-1}) or a resistance r (s cm^{-1}):

$$F/\Delta q = K/\Delta z = V = r^{-1}. \tag{15}$$

When the interface acts as a perfect sink, i.e., when $q=0$ at $z=0$,

$$F/q = V = r^{-1}$$

where q is the concentration of pollution at some convenient height. Tritiated water vapor and iodine-131, for example, are irreversibly absorbed at a snow surface. Frequently, however, the efficiency of interface exchange is poorly understood. The uptake of pollution by vegetation, for example, depends on whether the stomata are open or closed, while fluxes to and from the oceans are reduced by the presence of oil slicks.

In addition to interface characteristics, the flux depends on the rate at which the pollution is delivered to the surface, i.e., on the prevailing atmospheric turbulence regime, the integrated effect of which is given by the diffusivity K. When a strong ground-based inversion exists, for example, the pollution is trapped aloft and surface fluxes are small: thus a large value of $\Delta q/\Delta z$ does not necessarily imply a large surface flux. In a study of the uptake of sulphur by building stones, Braun and Wilson (1970) found that laboratory rates were much higher than those outdoors, indicating that atmospheric fluxes are limited by the rate of diffusion of SO_2 to surfaces. As might be expected also, the rates depended on the type of building stone, a coating of sodium carbonate resulting in 100% laboratory uptake over a 2 h period whereas most other types of surfaces yielded values of about 50%.

There are a number of experimental complexities in the estimation of surface depletion rates, particularly in irregular urban areas. Although reliable predictions are not yet possible for most pollutants, the investigator should at least recognize that the flux rather than the air concentration is of primary importance in depletion estimates.

6. The Influence of Pollution on Weather and Climate

The dispersion of pollution depends upon prevailing weather conditions. Conversely, weather and climate are influenced by the presence of pollution in the atmosphere, although the effects have not yet been well established. This lack of knowledge is due to incomplete observational networks (particularly on the regional and global scales), difficulties in designing meaningful experiments in the restless outdoor environment, and problems associated with feedback mechanisms, e.g., if pollution causes a global change in air temperature, the glacier and sea ice distributions will also change, resulting in a shift in storm tracks.

One of the most controversial questions is the effect of pollution on precipitation. Particulates act as condensation nuclei. Because the terrain surrounding a city or industrial area is usually non-uniform, however, it is often difficult to decide whether precipitation trends over decades, and non-uniform areal distributions, result from terrain or pollution effects. As early as 1929, Ashworth (1929) claimed to have found that there was 13% less rainfall on Sundays than on weekdays at Leicester, England, over a 10 yr period, presumably due to reduced emissions on Sundays. More recently Changnon (1968) described what he calls the LaPorte 'anomaly'. Figure 27 shows the

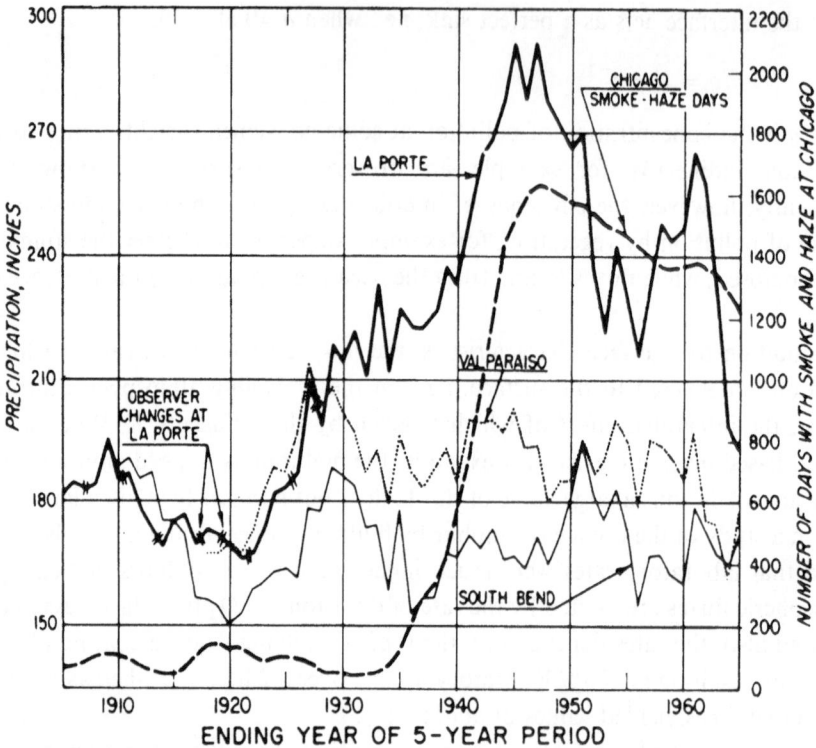

Fig. 27.   Precipitation values at several Indiana stations and haziness at Chicago, both plotted as 5 yr moving totals (Changon, 1968). Note the increase in precipitation at LaPorte which Changnon believes is caused by an increase in condensation nucleii from Chicago, as evidenced by the increase in haziness at the same time.

5 yr (1901–1965) moving-average precipitation totals at LaPorte, Indiana, located about 50 km east of Chicago. Included on the diagram are the precipitation values for two other Indiana stations, and the numbers of smoke-haze days in Chicago. Since 1925, there has been a 30 to 40% increase in precipitation at LaPorte with a concurrent rise in smokiness and steel production in Chicago. Changnon believes that there may be a cause-effect relation, due to the increase in condensation nuclei downwind from the industrial complex. Although Holzman and Thom (1970) challenge this hypothesis, suggesting that there may have been some kind of observational bias, Changnon (1970) marshals further evidence in support of the reality of the anomaly, including documentation of a secular rise in the frequencies of thunder and hail days and in crop-hail insurance losses in the LaPorte area.

    Ogden (1969) studied the data from 90 rainfall stations within 100 km of the Port Kembla steelworks in Australia. The stations had been established 15 yr before steelmaking began. Ogden could not find any evidence for a precipitation increase, despite the fact that the steelworks produce abundant condensation nuclei. Schaefer (1969), on the other hand, found not only large increases in condensation nuclei downwind of a number of United States cities (aircraft observations) but also ice crystal clouds and snow flurries in some cases. In order to determine whether this man-made modification triggers significant increases in cloudiness and precipitation, some carefully designed experiments on individual days are required. Even so, the same interpretive difficulties will arise as in cloud-seeding experiments: one cannot know whether the effect would still have occurred if there had been no nuclei source.

The effect of suspended particulate matter in changing the radiation balance of cities is well established. A number of studies have shown that the solar radiation is attenuated over urban areas, particularly during stagnation periods. There is speculation (e.g., Munn and Bolin, 1971) that large regional conglomerations of cities may produce anomalies in the average flow patterns, not only in the polluted areas but also in other parts of the world. There are great difficulties, however, in testing such a hypothesis.

On the global scale, there is public anxiety that the secular increases in CO_2 and particulate matter may change the world climate: the former causes the temperature to rise (the greenhouse effect) while the latter may cause cooling. The average tropospheric concentrations of CO_2 in the Northern Hemisphere are increasing at the rate of 0.7 ppm per year (Bolin and Bischof, 1970) while there is some evidence that turbidity is rising (McCormick and Ludwig, 1967). Whether these changes are affecting or will affect climate is a very difficult question to answer, however, because of feedback mechanisms. If increasing atmospheric CO_2 concentrations were to result in higher temperatures, for example, the ocean waters would become warmer, causing a decrease in CO_2 solubility and accelerating the rise in atmospheric concentrations. At the present time, the physical models of the general circulation are not sufficiently realistic to predict the influence of increasing pollution on the global climate.

7. Practical Aspects of Air Pollution Meteorology

A. METEOROLOGICAL INSTRUMENTS USED IN AIR POLLUTION

The meteorological elements most relevant in the study of dispersion of pollutants are the wind (both its mean and turbulent values) and the vertical temperature gradient. Other elements of secondary importance are precipitation, visibility, turbidity, and solar radiation. Table I lists the most important meteorological variables and the most common instruments that are used to measure them. For more detailed descriptions and illustrations see Middleton and Spilhaus (1953) and USDHEW (1967).

For maximum usefulness, data must be recorded in a reasonably compact and easily accessible form. One of the most reliable methods is to use the strip-chart recorder, in which an ink line is traced on a moving chart by a pen. The method is dependable, but subsequent extraction of data is time consuming. Another approach, and one which is practically imperative for measurements of turbulence, is to record the data on magnetic type or punched paper tape, which can be fed directly into electronic computers. In this case, an auxiliary strip-chart recorder is recommended in order that real-time quality control can be undertaken.

The siting of both meteorological and air pollution instruments requires consider-

TABLE I

Meteorological instruments used in air pollution

Element	Type of measurement	Instrument
Horizontal wind	Mean direction at a point Mean speed at a point Mean direction and speed at a point	Wind vane Rotating-cup anemometer Aerovane
Turbulence	Point	Bivane Hot-wire anemometer Sonic anemometer
Temperature	Point or gradients from a tower	Thermistor, Resistance thermo- meter (shielded and ventilated)
Humidity	Point	Psychrometer Hygrograph Hygristor (shielded and ventilated)
Wind	Variation with height	Tall tower Pilot balloon (pibal)
Wind Wind, temperature, humidity	Horizontal variation Variation with height	Tetroon Tall tower Rawinsonde Tethersonde Aircraft
Precipitation	Accumulation at the ground Rate at the ground	Standard rain gauge Tipping bucket rain gauge
Visibility	Path along the ground	Transmissometer
Turbidity	Direct sighting of the sun	Photometer
Solar Radiation	Ground	Pyrheliometer
Sunshine	Ground	Pyranometer Sunshine recorder Actinograph

able experience, particularly in built-up areas. The surface of the Earth is not homo-geneous as assumed in the models described in Section 2, and there is great difficulty in selecting 'representative' locations. This problem is so complex that the advice of a meteorologist should be sought in each case.

B. DIFFUSION FROM A CONTINUOUS POINT SOURCE

In most practical systems, Equation (3) is used to obtain the downwind ground level concentrations from a continuous point source. An averaging time of from 10 min to 1 hr is appropriate, with steady-state conditions being assumed. In order to construct an air pollution climatology, separate calculations are made for each hour, and subsequently combined to produce annual frequency distributions of pollution concentrations at various positions around the source.

The quantities to be evaluated are s_y, s_z, and h (effective stack height). The first two variables depend upon the turbulent structure of the flow and, over uniform surfaces, should be evaluated from direct observations of the cross-wind and vertical turbulent fluctuations, using the data-averaging techniques recommended by Pasquill (1961, 1962).

Frequently there is a need for order-of-magnitude estimates of s_y and s_z based on airport surface weather observations. Pasquill (1961, 1962) therefore developed a weather classification (Types A to F in Table II), which employs only simple measure-

TABLE II

The Pasquill stability classification

Surface wind speed (at 10 m) m/s	Day			Night	
	strong	insolation moderate	slight	thinly overcast or > 4/8 low cloud	< 3/8 Cloud
< 2	A	A–B	B	–	–
2–3	A–B	B	C	E	F
3–5	B	B–C	C	D	E
5–6	C	C–D	D	D	D
> 6	C	D	D	D	D

The neutral category, D, should be assumed for overcast conditions during day or night.

ments of surface wind speed, sunshine, and cloudiness. Types A, B, and C are associated with strong, moderate and slight instability, respectively, type D corresponds to a neutral lapse, and types E and F are associated with slight and moderate inversions, respectively.

For open countryside, the values of s_y and s_z to be associated with each weather classification type are given in Figures 28 and 29. These nomograms are based on observations obtained under ideal conditions during diffusion trials (Pasquill 1961, 1962; Gifford, 1961), and should not be used indiscriminantly in built-up areas or at other locations where there are abrupt changes in surface terrain. The advice of a meteorologist should be obtained in each case.

The other variable to be evaluated is the effective stack height, h, which depends upon exit velocity and gas temperature. Many plume-rise formulas exist, and there is merit in using several of them to obtain a range of values. Holland (1953) devised a formula for neutral conditions:

$$\Delta h = 1.5 \, (w_s d + 0.04 \, Q_H)/\bar{u}$$

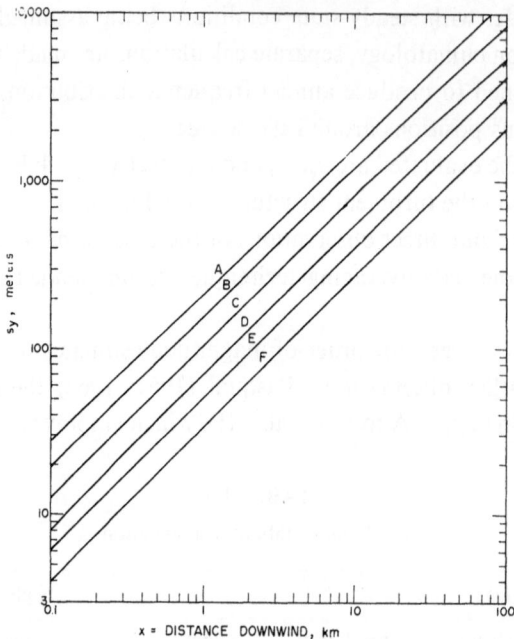

Fig. 28.   Standard deviation of the Gaussian distribution of pollutants in the cross wind direction
through a plume (USDHEW, 1967).

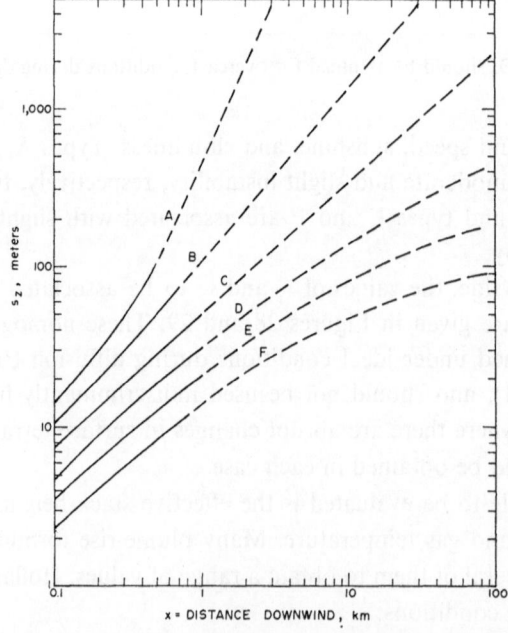

Fig. 29.   Similar to Figure 28 but for the Gaussian distribution in the vertical (USDHEW, 1967).

where the notation is explained in Section 2H. Holland recommended increasing Δh by 10 to 20% for instable conditions and decreasing it by the same amount during inversions. The CONCAWE formula (Brummage et al., 1966) is:

$$\Delta h = 0.175\, Q_H^{0.5}/\bar{u}^{0.75}.$$

Another estimate of plume rise can be obtained from the Moses and Carson (1968) formula (Equation (10)), the values of C_1 and C_2 depending upon stability:

	C_1	C_2
Unstable	3.47	10.53
Neutral	0.35	5.41
Stable	−1.04	4.58

The Tennessee Valley Authority report (1968) found that the CONCAWE formula showed the best agreement with observed rises of *hot* plumes. It was also found that, when more than one stack was operating, the plume rise was increased when the wind was along the line of stacks.

For *cold* plumes, the CONCAWE method is not recommended. Instead, Moses et al., (1964) found that the best results were obtained by using the Holland formula and doubling the computed plume rise.

C. AIR POLLUTION POTENTIAL

Regional air quality on a given day depends upon both emission strengths and atmospheric dispersion rates. The dispersive capability (or lack thereof) is called the air pollution *potential*, a quantity independent of source distributions or strengths. In non-industrialized regions whenever vertical and horizontal ventilation rates are unusually low, for example, pollution potential is high although air quality may be excellent.

As part of the general weather forecasting program for aviation and other interests, rawinsondes are released daily at 1200 GMT and 0001 GMT from a network of stations around the world, the density of the network in the United States being indicated by the dots in Figure 30. These observations can be used to obtain an approximate estimate of macro-scale daily pollution potential. Two mixing heights are calculated in the way shown schematically in Figure 31.

(1) *Afternoon maximum mixing height*: An adiabatic line is extended upward from the surface afternoon maximum temperature (forecast or observed) until it intersects the 1200 GMT temperature profile (1200 GMT is near sunrise over much of North America).

(2) *Urban morning mixing height*: Because rawinsonde stations are mainly at airport and rural locations and thus do not provide direct information on the urban mixed layer, a few degrees must be added to the rawinsonde surface temperature to simulate the effect of the city heat island. An adiabatic line is then extended upward until it intersects the 1200 GMT temperature profile, yielding an approximate estimate of the

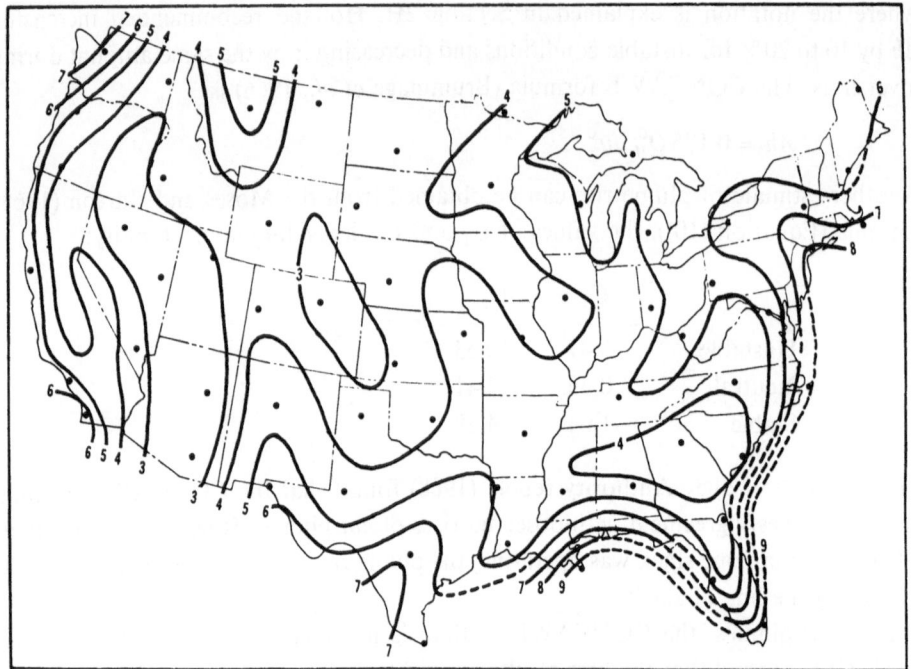

Fig. 30.   Isopleths of mean morning annual mixing height (10^2 m) from Holzworth (1970). The dots indicate the location of the radiosonde stations where mixing heights were computed.

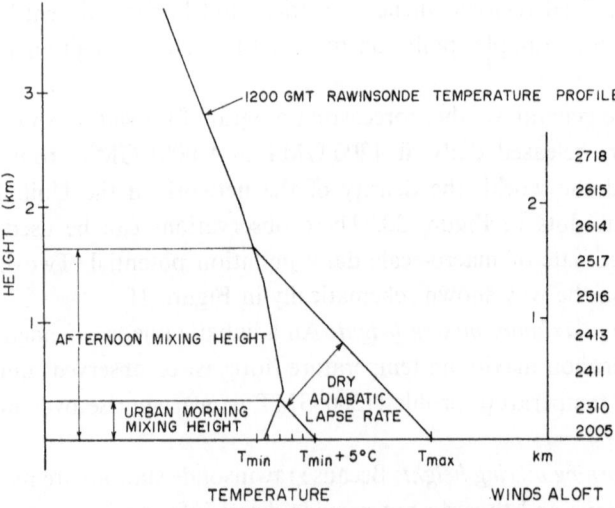

Fig. 31.   Computation of mixing heights (morning and afternoon) from a 1200 GMT radiosonde temperature profile. The mean transport wind within the mixing height can be found from scale of winds as a function of height on the right side of the figure. For instance: 2310 means a wind *from* 230° at 10 m s⁻¹ (USDHEW, 1967).

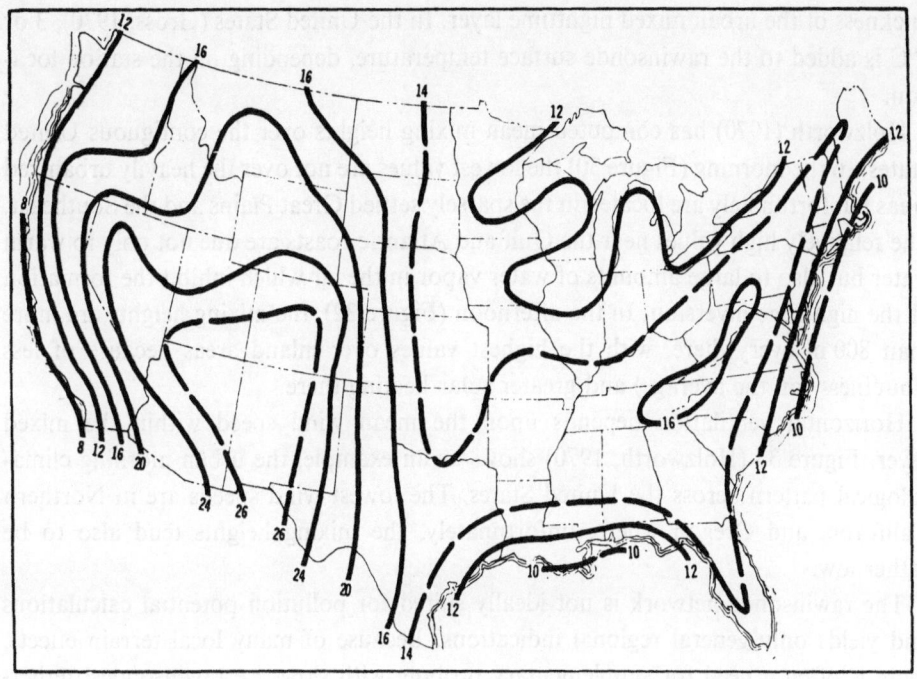

Fig. 32.   Isopleths of mean annual afternoon mixing heights (10^2 m), (Holzworth, 1970).

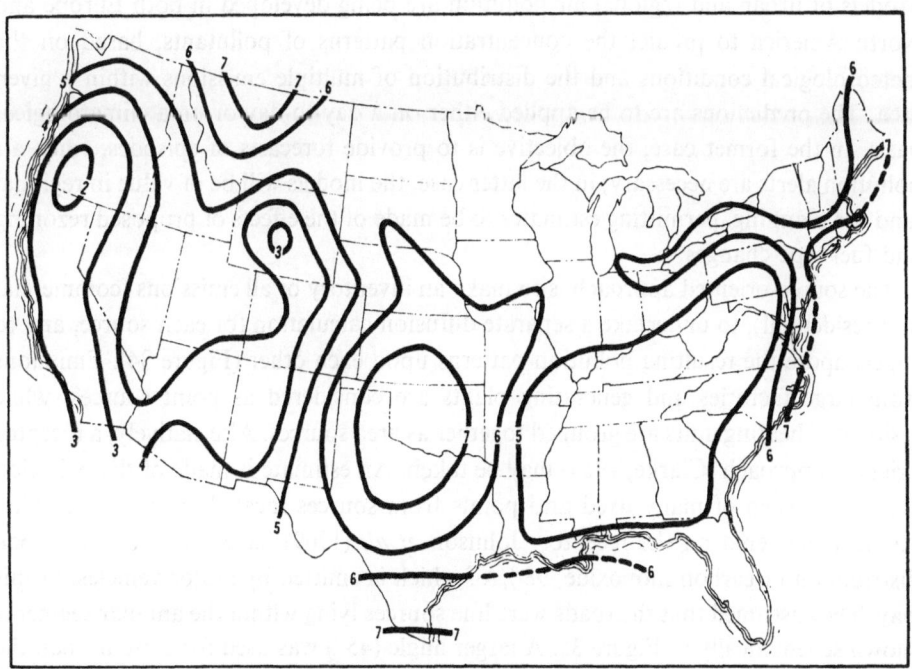

Fig. 33.   Isopleths of mean annual transport wind speed (m s^{-1}) averaged through the morning
mixing height (Figure 30), (Holzworth, 1970).

thickness of the urban mixed nighttime layer. In the United States (Gross, 1970), 3 or 5°C is added to the rawinsonde surface temperature, depending on the station location.

Holzworth (1970) has computed mean mixing heights over the contiguous United States. In the morning (Figure 30) the lowest values are not over the heavily urbanized areas but fortunately are located in the sparsely settled Great Plains and the Southeast. The relatively high values near the Gulf and Atlantic coasts are due not only to warm water but also to large amounts of water vapor in the air which inhibit the formation of the nighttime inversion. In the afternoon (Figure 32), the mixing heights are more than 800 m everywhere, with the highest values over inland areas because of less cloudiness (on the average) and greater solar heating there.

Horizontal ventilation depends upon the mean wind speed within the mixed layer. Figure 33 (Holzworth, 1970) shows as an example, the urban morning climatological pattern across the United States. The lowest wind speeds are in Northern California and Oregon where, unfortunately, the mixing heights tend also to be rather low.

The rawinsonde network is not ideally suited for pollution potential calculations and yields only general regional indications. Because of many local terrain effects, there is often a need for supplementary probing with slow-rise rawinsondes, tethersondes, and instrumented tall towers.

D. URBAN AND REGIONAL POLLUTION MODELS

Models of urban and regional air pollution are being developed in both Europe and North America to predict the concentration patterns of pollutants, based on the meteorological conditions and the distribution of multiple emissions within a given area. The predictions are to be applied either on a day-to-day or on a climatological basis. In the former case, the objective is to provide forecasts of episodes, when air pollution alerts are necessary; in the latter case, the models will be of value in regional land use planning, permitting estimates to be made of the effects of proposed rezoning and fuel-type changes.

The source-oriented approach is to make an inventory of all emissions (commercial and residential), to undertake a separate diffusion calculation for each source, and to superimpose the resulting pollution patterns upon each other (Figure 34). Emissions from large factories and generating plants are considered as point sources, while residential heating units are grouped together as area sources. Alternatively, a receptor oriented approach (Clarke, 1964) may be taken. An estimate is made of the pollution arriving at each of many fixed grid points from sources located in angular upwind segments centered on the receptor. Johnson et al. (1969) have modelled the urban distribution of carbon monoxide, 98% of which is emitted by motor vehicles, in this way. They assumed that the roads were line sources lying within the annular segments shown schematically in Figure 35. A larger angle (45°) was used for close-in than the angle ($22^1/_2$°) for distant sources, all emissions outside of these segments being assumed to make no contribution to the pollution at the grid point. Johnson et al. also incorpo-

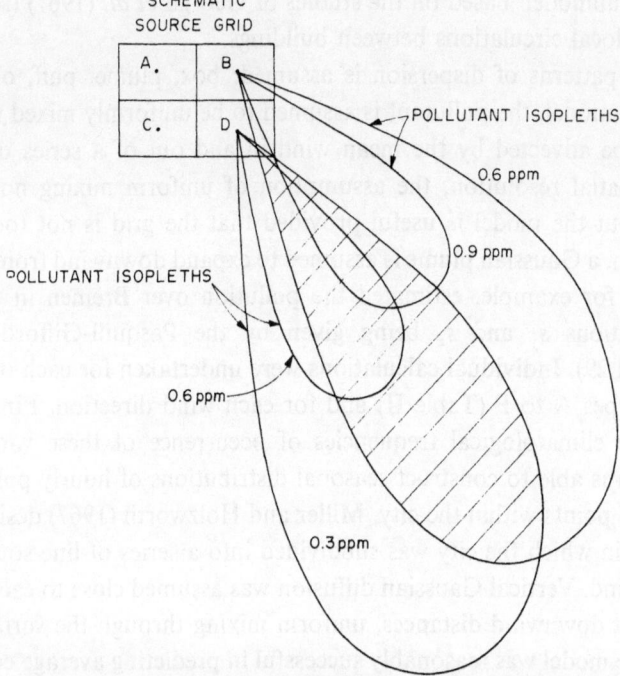

Fig. 34. A representation of a source-oriented pollution model. The patterns of pollution from all sources are superimposed and added to obtain the total concentration (Adapted from Moses, 1969).

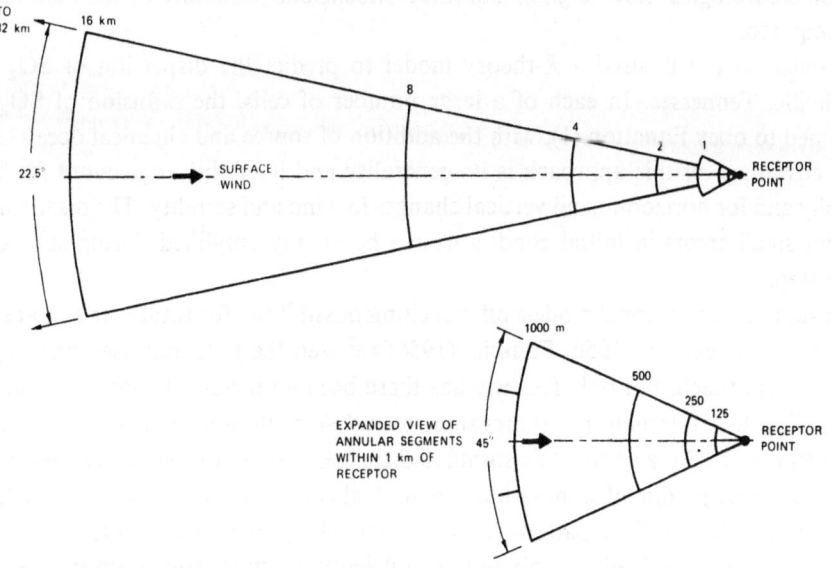

Fig. 35. A representation of a receptor-oriented pollution model (Johnson *et al.*, 1969). The receptor points are arrayed in a grid; the total pollution arriving at each receptor point is the sum of the contributions from all upwind annular segments. The spatial resolution increases towards the receptor.

rated a 'street submodel' based on the studies of Georgii *et al.* (1967) (see Figure 19) to account for local circulations between buildings.

One of four patterns of dispersion is assumed: box, plume, puff, or gradient-K. In the box-type model, the pollutant is assumed to be uniformly mixed to the mixing height and to be advected by the mean wind in and out of a series of boxes. The model lacks spatial resolution, the assumption of uniform mixing not being valid near sources, but the model is useful provided that the grid is not too fine. In the plume approach, a Gaussian plume is assumed to expand downwind from each source. Fortak (1969), for example, estimated the pollution over Bremen in this way, the standard deviations s_y and s_z being given by the Pasquill-Gifford nomograms (Figures 28 and 29). Individual calculations were undertaken for each of the Pasquill classification types A to F (Table II) and for each wind direction. Finally, from an estimate of the climatological frequencies of occurrence of these various weather types, Fortak was able to construct seasonal distributions of hourly pollution values at various fixed points within the city. Miller and Holzworth (1967) designed another kind of model in which the city was subdivided into a series of line sources, at right angles to the wind. Vertical Gaussian diffusion was assumed close to each line source, but at sufficient downwind distances, uniform mixing through the surface layer was postulated. The model was reasonably successful in predicting average concentrations over several American cities.

The Gaussian approach implies uniform wind and stability over the length of each plume, a condition often not met in a city. The puff model has therefore been tried, in Chicago, for example, by Roberts *et al.* (1969). When space and time variations are to be included in diffusion models, however, very detailed observations of the mesometeorological flow (e.g., of the three-dimensional structure of the lake breeze) are required.

Randerson (1970) used a K-theory model to predict the dispersion of SO_2 over Nashville, Tennessee. In each of a large number of cells, the diffusion of SO_2 was assumed to obey Equation (1), with the addition of source and chemical decay terms. The advantage of this approach is its generality and its ability to account for topography and for horizontal and vertical changes in wind and stability. The disadvantage is that small errors in initial conditions can be greatly amplified during subsequent time steps.

In summary, numerical models offer exciting possibilities for future air management programs. As early as 1956, Frenkiel (1956) realized the potential usefulness of this regional approach, but only recently has there been a serious attempt to develop his ideas. The first international symposium on urban pollution modelling was held in 1969 (Stern, 1971), and this has stimulated further research. The present position is that advances in digital computing are well ahead of those in our knowledge of mesometeorological flow patterns and of atmospheric pollution residence times. In addition, there is difficulty in obtaining sufficiently accurate source inventories. The diurnal cycle in emission strengths is not well documented in many areas, for example. Hilst (1969) has undertaken a useful sensitivity analysis of the Connecticut regional

SO$_2$ model, in which the values of the variables were deliberately changed to determine which elements were of most importance. His studies indicated that the predictions were relatively insensitive to the values of the cross-wind diffusion rate s_y and to random errors in source strengths; the forecasts were, however, strongly dependent on wind direction. This kind of study is to be encouraged, as an aid in designing urban and regional mesometeorological networks of stations.

References

Andersson, T.: 1969, 'Small-Scale Variations of the Contamination of Rain Caused by Washout from the Low Layers of the Atmosphere', *Tellus* **XXI**, 687–92.

Ashworth, J. R.: 1929, 'The Influence of Smoke and Hot Gases from Factory Chimneys on Rainfall', *Quart. J. Roy. Meteorol. Soc.* **55**, 341–50.

Batchelor, G. K.: 1949, 'Diffusion in a Field of Homogeneous Turbulence, I: Eulerian Analysis', *Australian J. Sci. Res.* **2**, 437–50.

Batchelor, G. K.: 1950, 'Application of the Similarity Theory of Turbulence to Atmospheric Diffusion', *Quart. J. Roy. Meteorol. Soc.* **76**, (328), 133–46.

Batchelor, G. K.: 1952, 'Diffusion in a Field of Homogeneous Turbulence, II: The Relative Motion of Particles', *Proc. Cambridge Phil. Soc.* **48**, 345–62.

Batchelor, G. K.: 1954, 'Heat Convection and Buoyancy Effects in Fluids', *Quart. J. Roy. Meteorol. Soc.* **80**, (345) 339–58.

Batchelor, G. K. and Townsend, A. A.: 1956, *Turbulent Diffusion, Surveys in Mechanics*, Cambridge University Press.

Beilke, S. and Georgii, H. W.: 1968, 'Investigation on the Incorporation of Sulphur-Dioxide into Fog- and Rain-Droplets', *Tellus* **XX**, 435–441.

Berg, O. T. G.: 1963, 'Investigation of Coalescence in Washout', Rep. 0780-01 (01). Aerojet-General Corp., Downey, California.

Bolin, B. and Bischof, W.: 1970, Variations of the Carbon Dioxide Content of the Atmosphere in the Northern Hemisphere', *Tellus* **XXII**, 431–42.

Braun, R. C. and Wilson, M. J. G.: 1970, 'The Removal of Atmospheric Sulphur by Building Stones', *Atmospheric Environ.* **4**, 371–8.

Briggs, G. A.: 1965, 'A Plume-Rise Model Compared with Observations', *J. Air Pollution Control Assoc.* **15**, 433–8.

Briggs, G. A.: 1969, *Plume Rise*, U.S. Atomic Energy Commission, Division of Technical Information, vi + 81 pp.

Brummage, K. G., et al.: 1966, *The Calculation of Atmospheric Dispersion from a Stack*, Stichting CONCAWE, The Hague, Netherlands, 57pp.

Byutner, E. K. and Gesina, F. A.: 1963, 'Effective Coefficient of Capture of Aerosol Particles by Rain and Cloud Droplets' (translated from *Trudy, Proc. Leningrad Hydrometeorol. Inst.* **15**).

Carson, J. E. and Moses, H.: 1969, 'The Validity of Several Plume Rise Formulas', *J. Air Pollution Control Assoc.* **19**, 862–6.

Chamberlain, A. C.: 1953, 'Aspects of Travel and Deposition of Aerosol and Vapour Clouds', British Report AERE-HP/R-1261.

Changnon, S. A.: 1968, 'Laporte Weather Anomaly, Fact or Fiction?', *Bull. Am. Meteorol. Soc.* **49**, 4–11.

Changnon, S. A.: 1970, 'The Laporte Precipitation Anomaly', *Bull. Am. Meteorol. Soc.* **51**, 337–42.

Clarke, J. F.: 1964, 'A Simple Diffusion Model for Calculating Point Concentrations from Multiple Sources', *APCA Journal* **14**, 347–52.

Clarke, J. F.: 1969, 'A Meteorological Analysis of Carbon Dioxide Concentrations Measured at a Rural Location', *Atmospheric Environ.* **3**, 375–83.

Csanady, G. T.: 1961, *Intern. J. Air Water Pollution* **4**, 47.

Csanady, G. T.: 1965, 'The Buoyant Motion Within a Hot Gas Plume in a Horizontal Wind', *J. Fluid Mech.* **22**, 225–39.

De Marrais, G. A.: 1961, 'Vertical Temperature Difference Observed over an Urban Area', *Bull. Am. Meteorol. Soc.* **42**, 548–54.

Facy, L.: 1962, 'Radioactive Precipitations and Fallout', in *Nuclear Radiation in Geophysics*, Springer-Verlag, Berlin.

Fick, A.: 1855, 'Über Diffusion', *Ann. Physik. Chem.* **2**, 94, 59–86.

Fortak, H. G.: 1969, 'Numerical Simulation of the Temporal and Spatial Distributions of Urban Air Pollution Concentrations', Paper presented at the NAPCA Symposium on Multiple Source Urban Diffusion Models, Chapel Hill, North Carolina.

Frenkiel, F. N.: 1956, 'Atmospheric Pollution and Zoning in an Urban Area', *Sci. Monthly* **82**, 194–203.

Georgii, H. W., Busch, E. and Weber, E.: 1967, 'Investigations of the Temporal and Spatial Distributions of the Emission Concentration of Carbon Monoxide in Frankfurt/Main', Reports of the Institute for Meteorology and Geophysics of the University of Frankfurt/Main.

Georgii, H. and Weber, E.: 1964, 'Investigations on Tropospheric Washout', Report AFCRL-64-816, Johann Wolfgang Goethe University.

Gifford, F.: 1955, 'A Simultaneous Lagrangian-Eulerian Turbulence Experiment', *Monthly Weather Rev.* **83**, 293.

Gifford, F.: 1959, 'Meteorology in Relation to Reactor Hazards and Site Evaluation', in *Proc. Sixth Lasagna Conf.* **II**, 7–19.

Gifford, F.: 1961, 'Use of Routine Meteorological Observations for Estimating Atmospheric Dispersion', *Nucl. Safety* **2**, 47–51.

Gifford, F.: 1962, 'Diffusion in the Surface Diabatic Layer', *J. Geophys. Res.* **67**, 3207–12.

Gross, E.: 1970, 'The National Air Pollution Potential Forecast Program', *ESSA Tech. Mem. WBTM NMC 47*, U.S. Dept. of Commerce.

Gunn, R. and Kinzer, G. D.: 1949, 'The Terminal Velocity of Fall for Water Droplets in Stagnant Air', *J. Meteorol.* **6**, 243–84.

Halitsky, J.: 1962, 'Diffusion of Vented Gas Around Buildings', *J. Air Pollution Control Assoc.* **12**, 74–80.

Hanna, S. R.: 1971, 'Simple Methods of Calculating Dispersion from Urban Area Sources', Paper presented at Conf. on Air Poll. Meteorol., Raleigh, N.C.

Hay, J. S. and Pasquill, F.: 1959, 'Diffusion from a Continuous Source in Relation to the Spectrum and Scale of Turbulence', *Atmospheric Diff. Air Pollution, Adv. Geophys.* **6**, 345, Academic Press.

Hilst, G. R.: 1969, 'The Sensitivities of Air Quality Predictions to Input Errors and Uncertainties', Paper presented at the NAPCA Symposium "Multiple Source Urban Diffusion Models", Chapel Hill, North Carolina.

Holland, J. Z.: 1953, 'A Meteorological Survey of the Oak Ridge Area', U.S. Atomic Energy Comm. Report ORO-99.

Holzman, B. G. and Thom, H. C. S.: 1970, 'The Laporte Precipitation Anomoly', *Bull. Am. Meteorol. Soc.* **51**, 335–7.

Holzworth, G. C.: 1970, 'Meteorological Potential for Urban Air Pollution in the Contiguous United States', Paper No. ME-20C presented at 2nd Internat. Clean Air Cong., Washington, D.C.

Johnson, W. B., Ludwig, F. L. and Moon, A. E.: 1969, 'Development of a Practical, Multipurpose Urban Diffusion Model for Carbon Monoxide', Paper presented at the NAPCA Symposium "Multiple-Source Urban Diffusion Models", Chapel Hill, North Carolina.

Junge, C. E.: 1963, *Air Chemistry and Radioactivity*, Academic Press Inc., New York.

Marshall, J. S. and Palmer, W. McK.: 1948, 'The Distribution of Raindrops with Size', *J. Meteorol.* **5**, 165–6.

Mason, B. J.: 1957, *The Physics of Clouds*, Oxford University Press, London.

May, F. G.: 1958, 'The Washout by Rain of Lycopodium Spores', British Report AERE-HP/R-2198.

McCormick, R. A. and Ludwig, J. H.: 1967, 'Climatic Modification by Atmospheric Aerosols', *Science* **156**, 1358–9.

McDonald, J. E.: 1963, 'Rain Washout of Partially Wettable Insoluble Particles', *J. Geophys. Res.* **68**, 4993–5003.

Middleton, W. E. K. and Spilhaus, A. F.: 1953, *Meteorological Instruments*, University of Toronto Press, xi + 286 pp.

Miller, M. E. and Holzworth, G. C.: 1967, 'An Atmospheric Diffusion Model for Metropolitan Areas', *J. Air Pollution Control Assoc.* **17**, 46–50.

Moroz, W. J.: 1967, 'A Lake Breeze on the Eastern Shore of Lake Michigan: Observations and Model', *J. Atmospheric Sci.* **24**, 337–55.

Morton, B. R., Taylor, G. I. and Turner, J. S.: 1956, 'Turbulent Gravitational Convection from Maintained and Instantaneous Sources', *Proc. Roy. Soc. (London)*, *Ser. A* **234**, 1–23.

Moses, H.: 1969, 'Mathematical Urban Air Pollution Models', Paper presented at 62nd Ann. Meeting of the APCA.

Moses, H. and Carson, J. E.: 1968, 'Stack Design Parameters Influencing Plume Rise', *J. Air Pollution Control Assoc.* **18**, 456–8.

Moses, H., Carson, J. E. and Strom, G. H.: 1964, 'Effects of Meteorological and Engineering Factors on Stack Plume Rise', *Nucl. Safety* **6**, 1–19.

Munn, R. E.: 1966, *Descriptive Micrometeorology*, Academic Press, New York, xiv + 245 pp.

Munn, R. E. and Bolin, B.: 1971, 'Global Air Pollution: Meteorological Aspects – A Survey', Submitted to *Atmospheric Env.*

Munn, R. E., Emslie, J. H. and Wilson, H. J.: 1963, 'A Preliminary Analysis of the Inversion Climatology of Southern Ontario', Rep. No. Tec-466, 13 pp. Meteorol. Branch, Toronto (CIR-3834).

Munn, R. E., Hirt, M. S. and Findlay, B. F.: 1969, 'A Climatological Study of the Urban Temperature Anomoly in the Lakeshore Environment at Toronto', *J. Appl. Meteorol.* **8**, 411–22.

Nakaya, U.: 1954, *Snow Crystals: Natural and Artificial*, Harvard University Press, Cambridge, Mass.

Nakaya, U. and Terada, T.: 1934, 'Simultaneous Observations of the Mass, Falling Velocity and Form of Individual Snow Crystals', *J. Fac. Sci. Hokkaido Univ.*, *Ser. 2* **1**, 191.

Ogden, T. L.: 1969, 'The Effect of Rainfall on a Large Steelworks', *J. Appl. Meteorol.* **8**, 585–91.

Oke, T. R. and East, C.: 1971, 'The Urban Boundary Layer in Montreal', *Boundary Layer Meteorology* **11**, 27–52.

Pasquill, F.: 1961, 'The Estimation of the Dispersion of Wind-Borne Material', *Meteorological Magazine* **90**, 33–49.

Pasquill, F.: 1962, *Atmospheric Diffusion*, Van Nostrand & Co.

Perkins, R. W. and Engelmann, R. J.: 1966, 'Trace Element and Trace Radionuclide Composition of Snow and Rain', USAEC Report BNWL-SA-650.

Peterson, K. R.: 1968, 'Continuous Point Source Plume Behaviour out to 160 Miles', *J. Appl. Meteorol.* **7**, 217–26.

Randerson, D.: 1970, 'A Numerical Experiment in Predicting the Transport of Sulphur Dioxide Through the Atmosphere', *Atmospheric Env.* **4**, 615–32.

Reiquam, H.: 1970a, 'An Atmospheric Transport and Accumulation Model for Airsheds', *Atmospheric Env.* **4**, 233–47.

Reiquam, H.: 1970b, 'Sulfur: Simulated Long-Range Transport in the Atmosphere', *Science* **170**, 318–20.

Roberts, J. J., Croke, E. J. and Kennedy, A. S.: 1969, 'An Urban Atmospheric Dispersion Model', Report from Argonne National Laboratory, Argonne, Illinois.

Roberts, O. F. T.: 1923, *Proc. Roy. Soc. (London)* **A104**, 640.

Schaefer, V.: 1969, 'The Inadvertent Modification of the Atmosphere by Air Pollution', *Bull. Am. Meteorol. Soc.* **50**, 199–206.

Schmidt, F. H. and Boer, J. H.: 1963, 'Local Circulation Around an Industrial Area', *Ber. Deut. Wetterdienst* **91**, 28–31.

Scorer, R. S.: 1959, 'The Rise of Bent-Over Hot Plumes', in *Advances in Geophysics*, Vol. **6**, Academic Press, Inc., New York.

Slawson, P. R. and Csanady, G. T.: 1967, 'On the Mean Path of Buoyant, Bent-Over Chimney Plumes', *J. Fluid Mech.* **28**, 311–22.

Stern, A. C. (Ed.): 1971, 'Proc. of the NAPCA Symposium on Multiple Source Urban Diffusion Models', Chapel Hill, N.C.

Summers, P. W.: 1964, 'An Urban Heat Island Model, Its Role in Air Pollution Problems, with Applications to Montreal', Paper presented at 1st Can. Conf. on Micrometeorol., Toronto, Ontario, Canada.

Sutton, O. G.: 1947, 'The Problem of Diffusion in the Lower Atmosphere', *Quart. J. Roy. Meteorol. Soc.*, *A* **135**, 143.

Sutton, O. G.: 1953, *Micrometeorology*, McGraw-Hill Book Company, New York, xii + 333 pp.

Taylor, G. I.: 1921, 'Diffusion by Continuous Movements', *Proc. Lond. Math. Soc.* **2**, 196–202.

Tennessee Valley Authority: 1968, 'Full Scale Study of Plume Rise at Large Electric Generating
    Stations', Tennessee Valley Authority, Div. of Health and Safety, Muscle Shoals, Alabama, 81 pp.
Tyson, P. D.: 1968, 'Nocturnal Local Winds in a Drakensburg Valley', *South African Geograph. J.*
    50, 15–32.
U.S. Atomic Energy Commission: 1968, '*Meteorology and Atomic Energy*', Report TID-24 190,
    x + 445 pp.
U.S. Dept. of Health, Education and Welfare: 1967, '*Meteorological Aspects of Air Pollution*'.

AIR POLLUTION – HUMAN HEALTH EFFECTS

RICHARD L. MASTERS

Head, Department of Aerospace and Environmental Medicine, Lovelace Foundation for Medical Education and Research, Albuquerque, N.M., U.S.A.

1. Introduction

Health, as used in this chapter, is defined not in the commonly used sense of 'good health', but rather by the definition of the World Health Organization: *Health is a state of complete physical, mental, and social well-being, not merely the absence of disease or infirmity.*

The discussion in this chapter is based on the above definition, and the viewpoint of preventive medicine, which concerns itself with the prevention of disease, prolongation of life, and promotion of physical and mental health and efficiency. As a further extension of preventive medicine, *environmental medicine* attempts to preserve and protect human life by a study of the effects of air pollution on human health. Environmental medicine, that branch of medicine dealing with the health effects of the interaction of the human body and the total environment, involves an interdisciplinary and multidisciplinary approach, and depends for its foundation upon the core sciences of physics, chemistry, and biology, as well as all of medical science.

Although a great deal of attention has been given to attempting to establish firm relationships between air pollution and health effects, the field still appears to some observers to be an admixture of facts, suppositions, and suspicions. Much of this difficulty arises from a confusion on the part of those unfamiliar with the health disciplines which have been brought to bear in the investigations. At the interface of this study of the many facets of air pollution there is brought together investigators with backgrounds in the biological, physical and chemical sciences. The usual certainty with which the physicist or chemist can resolve his questions is not consistently present in the biological sciences, and as one moves from pure biology toward clinical medicine and health effects on human beings, one also moves toward uncertainty. Hence, some scientists and engineers with their impressive array of problems met and solved, may have little patience with the more inexact areas of medicine. The public, too, has come to expect very firm answers from science. Hence, public impatience is also a problem.

Cause and effect relationships, frequently so firmly established by the rigorous disciplines of physics and chemistry and their subsciences, are not so easily established when we deal with humans. In experimentation we may subject the test animal to stimuli which may be uncomfortable, or pose a potential threat to his health or to his life. Obviously, humans often cannot be subjected to such stimuli. Hence, much of our information must be derived from experimentation on animals with inferences drawn from extrapolations we hope are valid, but which are, especially to the naive analyst, open to criticism. For those who insist on a firm cause and effect relationship, we must point out that one is unlikely to find such evidence here or in the voluminous literature accumulating on health effects of pollutants. Instead, there is presented a review of the accumulated evidence which stems largely from three sources, namely, (1) observed results of some episodes of human exposure, either accidentally or experimentally produced, (2) results of toxicologic studies on animals or humans, and (3) epidemiological analyses of effects based on observations and the statistical analyses of the various factors related on a time continuum thought to bear causal relationships to each other.

Increasing social concern and concern on the part of involved scientists over demonstrable and potential effects of air pollution derive from a number of factors. It is a well established fact that certain developments in science cannot occur until the various bits of knowledge necessary to produce such a development have become available. The development of newer techniques in toxicology, for example, is one of the elements which has led to increased awareness both on the part of scientists and in turn the public, of potential effects of air pollutants. Today pollutants can be measured in terms of extremely small quantities. Foreign substances within the body are detectable in extremely small quantities. The refinement of techniques of colorimetry and spectrophotometry has expanded our knowledge, leading to the understanding of more subtle effects than were previously measurable.

Further complicating the study of the health effects of air pollutants is the fact that, though we may be able to neatly categorize and define the various individual pollutants, such definitions and categorization may be artificial, since the interaction of pollutants may alter their effects on health, and the damage of one pollutant may not be identical from one individual to another. Exposure to certain pollutants can produce effects equal to the sum of each of the pollutants acting independently, producing what is referred to as an *additive* effect. Some pollutants may work against each other with the resultant effect less than either would or could produce alone, an *antagonistic* effect. Still other combinations of pollutants when acting together may produce results greater than the sum of the individual pollutants, producing an enhanced or *synergistic* effect. *Potentiation* is the enhanced response of a combination of two or more substances over and above that of expected effects, but in which one of the substances alone would be either inactive or would produce a response of a different kind than that of the combination. (An example of potentiation is the enhanced effect of the combination of sodium chloride aerosol and sulfur dioxide (SO_2) gas. Here, the sodium chloride in the concentration used has no action alone.)

As research is proceeding, more and more information is evolving to link air pollution to health effects. Also, air pollution is not the only dragon to be slain. Air pollution does not act alone. Other factors such as water pollution, food contamination, radiation, to mention a few, all may act singly or in combination to add to the total body burden and to the possible production of detrimental effects on health.

However, man is, with air pollution perhaps more than any other form of environmental insult, the involuntary recipient. We may avoid polluted water or contaminated food, and we can do a great deal to control or contain radiation, but breathing is not controlled voluntarily, and we must breathe to live.

As the expansion of scientific studies of air pollution effects continues, some of the elements which are now 'unknowns' will become 'knowns'. We may indeed be required to substantially revise our estimates of possible effects of air pollution on humans and animals and other life forms. We will discover effects which today are completely unknown. As the scientific capabilities of chemistry and physics are brought to bear in toxicology, we have already surpassed our capabilities to detect effects at the measurable levels now possible for the various pollutants. Often in the past the study of air pollution effects has been limited to those instances of acute episodes with high concentrations of pollutants. Now we are more concerned with the chronic low level effects. To adequately study chronic low level effects in man it is sometimes necessary to consider the entire life span of a man. Is it fair to wait the entire life span of man to determine possible effects? Can we shorten this span of waiting? Is it not now the direct responsibility of scientists to adopt an attitude of attempting to protect humans on the basis of good presumptive probable evidence rather than pure cause-effect relationships? No one would encourage scientists to come to unwarranted conclusions; however, scientists can no longer find themselves in the untenable position of utilizing an unreasonable requirement for absolute proof as a convenient excuse for social irresponsibility.

2. Basic Anatomy and Physiology

Access of all materials to the body is necessarily through some type of membrane, be it the skin or one of the membranes lining the tracts of the body or forming the external surface of the eye. Target organs upon which air pollutants must act to exert their effect or to gain entry to the body where they may exert effects are the skin, mucous membranes, conjunctivae of the eye, the linings of the respiratory tract and lungs, and the linings of the digestive tract.

A. THE SKIN

The integument, or skin, is a remarkable organ which is specially adapted to containing the body and at the same time protecting it from environmental agents which would be harmful. Gases can penetrate the skin freely and liquids penetrate less freely. Lipid- and water-soluble compounds can penetrate, but solids insoluble in water or lipid cannot penetrate the skin. Nevertheless, the concentrations of pollutants usually found in the atmosphere are of little consequence. Naturally, particulates will cause the skin to become dirty, and this in turn can reduce the efficiency of the skin function, such as sweating to control body temperature. But the argument here is academic. The mucous membranes of the digestive tract in the mouth can, in some instances of high air pollution, literally enable the recipient to 'taste' air pollution, but usually only in episodes of intense pollution, such as may be found in industrial operations, very close to the source of production of the contaminant. The tongue is specifically equipped to detect acid, sour, bitter, salty and sweet elements. Odor and taste are inseparably intertwined body functions, and may be the detector mechanisms for air pollutants.

B. THE DIGESTIVE SYSTEM

Air pollutants gain access to the digestive system usually in such low concentrations as to be of no practical importance. A small amount of air is swallowed in the course of everyday activities such as eating, gum chewing, etc. Air pollutants which enter the digestive tract in this manner have not been demonstrated to have any harmful effects, since the low concentrations are further diluted by the secretions of the tract and by food matter and water. Material exerted from the respiratory tract by means of the respiratory cleansing mechanisms may be swallowed, thus gaining access to the digestive tract. Exceedingly toxic materials may present a serious hazard, and may gain entry in this manner when they would have no chance of being taken voluntarily into the digestive tract. Some pollutants pass through the digestive tract unaltered, some unabsorbed. Absorption is an active process, requiring the interaction of the membranes lining the digestive tract and that substance which is to be absorbed, such as a food element. Chemical reactions in the digestive tract prepare and digest food for absorption. Elements which are not of nutritive value to the body are often not absorbed at all, passing directly through the tract to become excreted in the feces.

C. THE RESPIRATORY SYSTEM

The respiratory system is the organ system which is most frequently assaulted by air pollutants. Basically, its components are the nose and nasal passages, the trachea, the bronchial tubes and the lungs themselves. Although we may arbitrarily divide the respiratory tract into upper and lower portions, or by subdivisions such as listed above, it must be remembered that it functions as a physiologic unit, serving to warm and humidify the ventilated air. Its ultimate function is the gaseous exchange between the inhaled air and the blood gases. Figure 1 illustrates the basic anatomical sub-divisions of the respiratory system. Reference to this figure will assist in visualizing the pathway of air in and out of the body.

As the air is drawn into the nasal passages, it passes through a system of baffles, where larger particles impinge on the mucous membranes, and are trapped there to be removed by the cleansing mechanism. Microscopically, the mucous membranes have a cell structure consisting of cilia, which are small hair-like projections. The cilia beat in an upward direction, sweeping foreign particles upward and out of the tract. Coughing and normal breathing movements assist in moving the larger accumulations.

From the nasal passages and pharynx the air passes through the larynx and trachea. Impingement of particles is still possible and indeed common in this portion of the tract as well. The trachea branches into the right and left main stem bronchi, and a pattern of dichotomous and monopodial division is repeated with decreasing cross-sectional diameters of the tubes until the level of the respiratory bronchioles is reached. Here, even more extensive branching takes place, giving rise to the alveolar ducts, alveolar sacs, and the alveoli themselves. It is in the alveoli that the exchange of gases between the ventilated air and the blood takes place; and in the alveolar sacs the walls are only a few molecules thick. There is approximately 70 to 80 m^2 of alveolar surface area for the exchange of gases. The alveoli are maintained as essentially sterile and free of foreign matter with a basal alveolar ventilation of approximately 4 liters a minute.

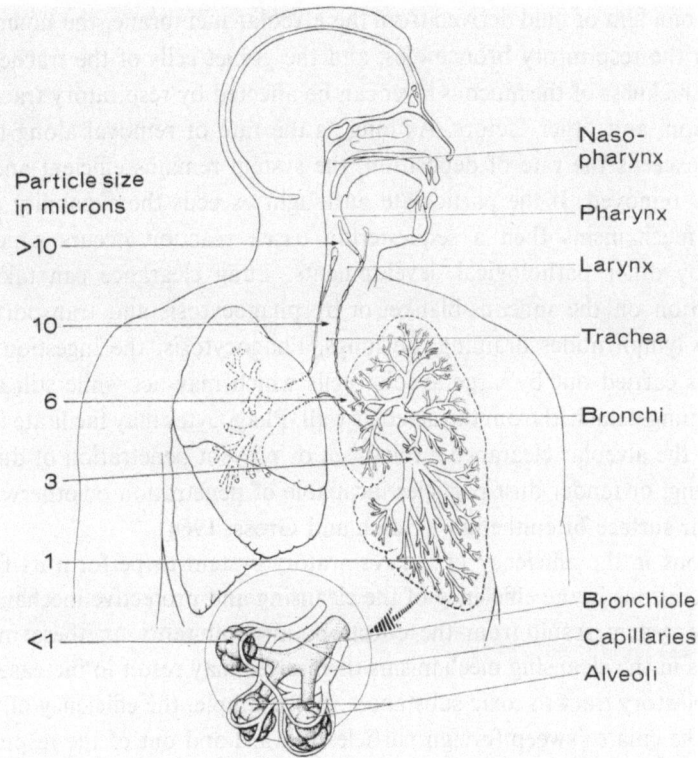

Fig. 1.   Schematic illustration of the respiratory tract and its anatomic subdivisions. Indicated on the left are the estimated deposition sites for indicated particle sizes. For example, particles of greater than 10 μ in diam are rarely encountered beyond the larynx. (Based on data from Dautrebande and Walkenhorst, 1964; Davies, 1964; Hatch and Gross, 1964; Goddard et al., 1968.)

Dautrabande and Walkenhorst (1964) in their analysis of the deposition of micro-aerosols in human lungs found that the deposition of particles of small size does not take place until the particles reach the alveolar sacs. The alveolar deposition of the particles is apparently not significantly influenced by duration of dust laden air inhalation or breathing pause. Minimum retention size was found to be about 0.6 μ for coal dust, 0.2 to 0.3 μ for insoluble submicroscopic particles, and 0.05 to 0.1 μ

for dry sodium chloride crystals. The maximum size of particles found in deep alveolar air has never exceeded 1 μ as observed by electron microscopy. The most frequently observed sizes were from 0.05 to 0.4 μ.

Pulmonary clearance mechanisms are important protective elements of the respiratory system. Clearance of particles from the lungs involves the transportation of particles from the site of deposition within the alveoli to the terminal bronchioles and then with the moving blanket of mucus which covers the respiratory tract, up to the throat. At this point they are commonly swallowed. The beating motion of the ciliated cells lining the respiratory tract results in a propulsion of a normal mucinous fluid layer at an average rate of about 13 to 14 mm/min. The mucous blanket itself is composed of a film of fluid derived from the alveolar membrane, the mucus secreting cells lining the respiratory bronchioles, and the goblet cells of the tracheobronchial tract. The thickness of the mucous layer can be affected by respiratory tract irritation, inflammation, and other factors. As long as the rate of removal along the airways equals or exceeds the rate of deposition, the system remains efficient and materials are rapidly removed. If the particulate onslaught exceeds the capability of the lung clearance mechanisms then a sequestering tissue reaction occurs which may be followed by other pathological developments. Lung clearance can take place by transportation on the mucous blanket or by phagocytosis and transportation of a particle to lymph nodes draining the lungs. Phagocytosis, the ingestion of foreign material, is carried out by large aveolar cells which may act while still attached or after becoming detached from the alveolar wall. Phagocytes may facilitate the removal of dust by the alveolar clearance mechanism or prevent penetration of dust particles into the lung, or render dust particles incapable of penetration or otherwise injuring the alveolar surface of epithelium (Hatch and Gross, 1964).

Alterations in the efficiency of the respiratory system to perform its function, as well as alterations in the efficiency of the cleansing and protective mechanisms of the respiratory system result from the effects of air pollutants on these mechanisms. Alterations in the cleansing mechanisms themselves may result in increased exposure of the respiratory tract to toxic substances. For example, the efficiency of the beating action of the cilia to sweep foreign particles upward and out of the respiratory tract may be affected by certain air pollutants. A decrease in the efficiency or a complete cessation of activity of the cilia can cause pollutants to become fixed at a particular point in the tract, thus increasing the likelihood of their creating a toxic effect. Also, the slower the beating action of the cilia, the slower the upward sweep of foreign material. In time, this would increase the residence of foreign matter, thus also increasing its ability to exert a toxic effect on the respiratory tract lining. Sulfur dioxide has an effect on the ciliary action of decreasing or stopping the action, depending upon the concentration of sulfur dioxide used in the exposure. It has been found that SO_2 administered prior to the challenge of disease producing microbes in guinea pigs resulted both in a decrease in the action of the cilia and an increase in the likelihood of infection resulting from the disease organisms (Ehrlich, 1963). The mucous film may also be thickened as a result of SO_2 irritation.

3. Basic Toxicology

Toxicology is the study of the harmful actions of chemicals on biologic tissue, and involves an understanding of chemical reactions and interactions and of biologic mechanisms (Loomis, 1968). Toxicology is divided into environmental, economic, and forensic branches.

Environmental toxicology is the study of the harmful effects of chemicals encountered by man through air and water pollution, industrial exposures, chemicals added to animal feed, and food additives.

One of the most fundamental observations to be made respecting any biologic effect of a chemical agent is the relationship that exists between the dose and the response. The single factor that determines the degree of harmfulness of a toxic compound is the dose. What may be harmful to one biologic specimen may not be harmful to another, and, in fact, can even be desirable. Hence, toxicity is a relative property of chemicals and can be directly or indirectly undesirable as far as man is concerned. The study of dose-response relationships requires a categorization of toxicity. The following table derived from Loomis (1968) demonstrates categories of toxicity:

Toxicity classification	Dose-response level
Extremely toxic	1 mg/kg or less
Highly toxic	1–50 mg/kg
Moderately toxic	50–500 mg/kg
Slightly toxic	0.5–5 g/kg
Practically non-toxic	5–15 g/kg
Relatively harmless	15 g/kg or more

Tolerance is the ability of an organism to show a decreasing response to a specific dose of a chemical with repeated exposure. That is, there is a lesser response to a specific dose than that which occurred on prior occasions from the same dose. Tolerance is known to occur with repeated exposure to habit-forming drugs, and can develop after repeated exposure to environmental agents.

A chemical agent must come in contact with the reactant chemical at its location in the biologic system if a chemical reaction or a physical effect is to occur. The reaction between two chemicals takes place readily only if certain criteria are met. The chemicals must be suitably selected and placed in physical contact with each other so that a reaction will take place. In addition, chemicals must be soluble to some degree in some vehicle. Unless the product of reaction is removed, the reaction does not go to completion; equilibrium results. These general principles as applied to chemical reactions usually apply in biologic systems also.

There are two categories of chemical factors which are generally felt to influence toxicity. The first consists of those chemical and physiochemical properties of compounds which individually and collectively determine the ability of the compounds to pass across biologic membranes. These regulate, in a sense, the transport of the chemical throughout the biologic system. A second category involves the chemical structure of compounds which enables them to produce specific actions on the tissues and which may be susceptible to the transformation by mechanisms present in the biologic system. For example, compounds which are less toxic than the parent compounds or products possessing greater toxicity than the parent compound may be formed. Specific chemical reactions, resulting in generalized destruction, can be produced by any chemical which is sufficiently soluble in tissue fluids to gain access to the cells in the sufficient concentration. Many of the chemicals of interest to the toxicologist involve selective chemical action, whereby drugs or chemicals produce their harmful effects at specific sites and concentrations, far below the concentrations normally necessary to produce overwhelming destruction of cells. Further factors influencing the transportation or translocation of chemicals within the body are ionization and the solubility of chemicals in its fluids. Biotransformation mechanisms or metabolic transformations involve oxidation and reduction reactions, hydrolysis, acetylation and conjugation reactions which are catalyzed by enzyme systems. The determination of the toxicity of any compound which undergoes metabolic trans- formation must involve the study of the toxicity of the parent compound and of its satellites. Some examples of chemicals whose toxicity is increased from metabolic conversion are: methanol is metabolically converted to formaldehyde in the body; nitrobenzene is converted to nitrosobenzene and phenylhydroxylamine; acetanalid is converted to analine; codeine is converted to morphine. The metabolic transformation of parathion to paraoxon in insects is quite important and is the mechanism whereby parathion becomes lethal to insects.

It is important to understand the dynamics of deposition and removal of particulate matter from the respiratory tract. Particulate matter is categorized into dusts, fumes, mists and fogs. *Dusts* are solid particulates and are generated by such operations as grinding or crushing of such substances as rock, metal, wood or grain. Particle diam- eters are variable. *Fumes* are solid particles generated by condensation from the gaseous state as from volatilization of molten metals. Particle diameters of fumes are less than one micron. *Mists* are suspended liquids in droplets generated by condensa- tion from the gaseous to the liquid state as by atomizing or splashing. Examples are oil mists, spray paint, etc. *Fogs* are liquid particles of condensates with particle sizes larger than mists, usually greater than 10 μ. Supersaturated water vapor in the air is a good example. *Gases* are formless fluids occupying completely the space of an enclosure and which can be changed to the liquid or solid state by the combined effects of in- creased pressure and increased temperature. Carbon monoxide and hydrogen sulfide are examples of gases. *Aerosols* are dispersions of particulates in a gaseous medium while *smokes* are gaseous products of combustion. *Vapors* are the gaseous forms of substances which are normally in the liquid or solid state and which can be trans- formed to these states either by increasing the pressure or decreasing the temperature. An example of a vapor is gasoline.

Special consideration is necessary in dealing with the concept of dose-response relationships when applied to aerosol and particulate inhalation. Atmospheric concentration and rate of breathing are not sufficient information to allow the drawing of conclusions regarding the size of dose. In a detailed treatise on the subject, Hatch and Gross (1964) point out that it is necessary, in order to determine a quantitative dose-response relationship at the critical site, to estimate the amount of inhaled aerosol initially deposited and to determine where it is deposited in the respiratory tract and lungs, and what fraction of the retained material actually reaches the critical site of action in the lungs or other parts of the body to produce the effect, damage, etc. Obviously, the exercise is not a simple one, and requires considerable knowledge of physiology and physics. Aerosol deposition within the respiratory system is affected by a number of physical factors, including the size of the particle, the air velocities and transit times of air within the system, obstructions and baffles encountered, changing direction of air flow, and the forces necessary to displace particles sufficiently to impinge or precipitate them onto the surfaces of the respiratory tract lining. The particle diameter which is believed to cause the most injurious response is less than 1μ. Larger particles either do not remain suspended in the air sufficiently long to be inhaled or if they are inhaled cannot negotiate the baffle system of the upper respiratory tract as described above. Smaller particles are also deposited in a much larger percentage of the exposure concentration than larger particles. Smaller particles appear to be less readily removed from the lungs thus adding additional dosage and residence time. Particles of high density because of their greater mass and consequent inertia impact on the walls of the upper respiratory tract. Environmental temperature modifies the toxic response of inhaled materials with high as well as low temperatures increasing the undesirable effect. The absorption and retention of gases and vapors depend very much on the solubility of the gas in the environment of the respiratory tract. Ammonia or SO_2 will be rapidly soluble and will be retained in the upper respiratory tract with very little reaching the pulmonary alveoli. Of course, this is dependent upon concentration of the gas. On the other hand, ozone and carbon disulfide which are insoluble will reach the alveoli and very little will be absorbed in the upper respiratory tract.

From the considerations of the elementary anatomy, physiology and toxicology discussed above, it can be seen that a most remarkable characteristic of living organisms, including man, is the ability to adapt to hostile environments. After gaining entrance to the body, materials in the environment are subjected immediately to a series of defense mechanisms which protect the body from the onslaught of the foreign chemicals. If the order of magnitude and toxicity of the assaulting pollutants is not great enough to overcome the adaptive and protective mechanisms of the organism, then the organism may survive for unlimited periods within a polluted environment. This simplistic evaluation is not sufficient, however, to account for possible ultimate influences on the survival of species. Extinction is not a unique phenomenon in natural history, and is largely the termination of an unsuccessful attempt to adapt to an environmental encumbrance.

4. Methodology of the Study of Health Effects

An impressive body of evidence has been accumulated to correlate air pollution with (1) increased mortality from cardiorespiratory causes, (2) increased susceptibility to respiratory disease, and (3) interference with normal respiratory function. This evidence comes primarily from three types of investigations. First, there are retrospective statistical studies of past morbidity and mortality which correlate geographic location, meteorologic phenomena, and other physical and demographic factors with air pollution. Second, prospective epidemiological studies of relations of morbidity, mortality and respiratory functions to variations in air pollution are underway. Third, there are laboratory studies of response by animals and by humans, in some cases, to exposure to various pollutants singly or in combination. A few examples of the results of these several types of studies will be of interest.

Death rates from cardiorespiratory causes in the United States are greater in urban than in rural areas, and, in general, increase with city size as does air pollution. It has been demonstrated that the emphysema death rate, when sex and age are accounted for, increases from the lowest in rural areas to intermediate in areas surrounding major cities, to its highest level within the central portions of immediate metropolitan areas. Within the last few years this urban-rural difference, referred to as the *urban effect* has also been shown for mortality of infants less than one year of age and is probably accounted for by respiratory causes of death.

Sterling *et al.* (1969), in a study of urban morbidity and air pollution in Los Angeles, found a significant correlation between pollution and fluctuation of rate of admissions to hospitals for disease categories including allergic disorders, inflammatory diseases of the eye, acute upper respiratory infections, influenza, bronchitis and certain cardiovascular diseases.

Hodgson (1970) conducted a study of the short-term effects of air pollution on mortality in New York City, analyzing data on mortality, air pollution, and meteorological factors during the period November 1962 through May 1965. He concluded that the level of respiratory and heart disease mortality over time is quite significantly related to environmental conditions, and that 73 % of the variation in mortality from heart and respiratory diseases was explained by concurrent variations of air pollution and temperature, with pollution being the more important explanatory variable. The influence of the environment on respiratory and heart disease mortality was not strictly limited to those over 65. The suggestion that the increase in mortality represented bunching together of deaths of persons already ill at the time of increased air pollution will not hold, since if the total effect of air pollution was to redistribute deaths within a short time interval, there would be no statistically significant increase in the relation between monthly mortality and air pollution. The categories of mortality related to air pollution were found to be pneumonia, vascular lesions affecting the central nervous system, arteriosclerotic heart diseases (including coronary disease and hypertensive heart disease), and other diseases of the heart. Hodgson points out that the linear relationship between mortality and environmental factors implies that a unit change in the level of air pollution induces just as great an increase in mortality at low levels of pollution as at higher levels.

Glasser and Greenburg (1971) estimated that the increase in the mean number of daily deaths in New York City, during the period 1960–1964, was ten to twenty deaths per day when mean SO_2 levels of 0.40 ppm or more were compared with levels of 0.20 ppm or less.

In Great Britain there is a very high incidence of bronchitis, higher than in the United States, for example. Fairbairn and Reid (1958) reported that the incidence of severe chronic bronchitis causing disability or death among postmen and deaths among middle-aged men and women in 37 different areas of Great Britain were related to the frequency of winter fog, and this presumably to the level of air pollution. Pneumonia mortality was also significantly correlated with fog frequency. Sickness absences from bronchitis and pneumonia were higher among the postmen who worked outdoors in areas where there was a high fog index than in the postal clerks who worked indoors in the same areas. Further, bronchitic postmen who had worked in areas having the highest SO_2 and particulate levels showed far more frequent absences due to bronchitis, pleurisy and pneumonia than bronchitic postmen working in less polluted areas. Other investigators have reported close associations between clinical conditions of patients with chronic bronchitis and emphysema and the concentration of atmospheric pollutants. The standardized mortality ratios for bronchitis have been significantly related to the average amount of undissolved deposits in the county and urban areas and to smoke concentration in the county area. There is also a significant correlation between mortality from bronchitis and the average concentration of sulfur dioxide in the county boroughs of England and Wales. The bronchitis mortality rate is higher, especially in males, in urban than in rural areas of Great Britain. Possible criticisms of the applicability of these studies in England to the United States or to other areas of the world result from the fact that the type of air pollution found in London and in other British metropolitan areas, where coal is widely used for heating, may not be similar to that found in other parts of the world. For example, in many cities of the United States, petroleum products are the major source of air pollution.

Of great importance to health authorities is the effect of repeated exposure to low levels of contaminants such as exist in urban communities or in special localized areas. Epidemiologic studies have been buttressed by clinical and physiological studies of persons experimentally or accidentally exposed to various degrees of air pollution, by studies of industrial workers who are exposed to chemicals in their work environment, and by experiments on animals. All of these methodologies are subject to certain limitations. Morbidity and mortality rates are affected by so many variable factors, in addition to air pollution, such as social condition, economic status, occupation, cigarette smoking, the presence of other diseases, etc., that it is almost impossible to determine the specific etiologic role of any one factor. It is also difficult to differentiate a direct effect of air pollution in initiating a disease from a possible aggravating effect of pre-existing disease. Epidemiologic analyses suffer from a lack of accurate information about the character and extent of atmospheric pollution. Estimates of air pollution based on visual judgement, or soot fall measurements, are open to serious question. Sedimentation samples may be composed of large particles or conglomerates which do not reach the lungs and may settle too rapidly to even affect the nasal tissue. The relationships established between industrial exposure and effects may not be applicable to the general population or especially to more susceptible members of the population.

Animal experiments are subject to a number of drawbacks. There is no question that the 'perfect' model for the study of man would be man himself, but ethical considerations make it impossible to use man in exposures where irreparable damage may be done. Hence, the use of animal models continues and is of vital interest. The criteria for animal models must be such that they approach with some reasonable degree of certitude the reactions which will be found in the human body. Further, exposures must relate the differences in anatomical structure and physiologic function between the systems of the animal and man. Also, exposures must be related to the relative sizes of the animal and man. The animals to be used as models must be capable of developing one or more of the diseases or pathological conditions which we feel may be associated with the exposure in man. Further, the animals must demonstrate a dose-response relationship which is not unlike that of man to similar toxic materials. Laboratory studies involving exposure of the animals, and in a few instances, humans, to controlled concentrations of irritant pollutants agree generally with the results of the epidemiological studies. For example, exposure to such gases as ozone and SO_2 under laboratory conditions results in decreased respiratory rate and increased total pulmonary flow resistance. Animals exposed to irradiated automobile exhausts also demonstrate these changes. Exposure to irritant pollutant gases such as ozone and SO_2 slows down or stops the beating motion of cilia, thus interfering with this lung clearing mechanism. This effect has been demonstrated in the laboratory both in excised tissue taken from the respiratory tract and also from intact animals. These observations may help to explain the mechanism involved in the greater susceptibility to respiratory disease of laboratory animals pre-exposed to irritant gases and pollutants.

All of these examples cited, of course, provide associative evidence only and cannot be taken as *proving* a cause-effect relationship between air pollution and effects upon health. It may well be that such cause and effect relationships can never be fully proved and that further studies will only add to the accumulating body of 'circumstantial evidence'. It now appears likely that the doctrine of specific etiology of diseases, which states that each disease has a single and specific cause, and which has been so fruitful in advancing health practices in communicable disease control, may not apply in chronic disease causation with low levels of pollutants. Thus, in speaking of chronic diseases, Dubos (1959) has stated that "the search for *the* cause may be a hopeless pursuit because most disease states are the indirect outcome of a constellation of circumstances rather than the direct result of single determinant factors".

5. Some Disease Risks from Inhalation of Specific Aerosols

A. PNEUMOCONIOSES

Pneumoconioses are pathologic conditions caused by reaction of the lungs to fine dust particles. Examples are silicosis, asbestosis, coal workers' pneumoconiosis, and talc miners' pneumoconiosis. Pneumoconioses may be benign, causing little or no ill effects, or they may range to excessively serious respiratory damage and fibrosis of the lungs. Combined with other insults such as tuberculosis, pneumoconioses may become rapidly progressing disabling diseases. Variations in the composition and particulate nature of the causative agents of the various pneumoconioses are of considerable importance in the likelihood of the development of disease. For example, high crystalline silicone dioxide content of inhaled dust is known to increase the risk of silicosis. Several of these diseases will be discussed later in this chapter.

B. ALLERGIC CONDITIONS

Because specific sensitizing substances can be inhaled, deposited and retained on the respiratory tract surface, it is appropriate to include allergic reactions among those disorders which may result from inhalation of air pollutants. *Allergy* is the tendency of an organism to manifest abnormal altered reactivity to a specific chemical grouping acquired in the course of prior exposure to that grouping (Pitts and Metcalf, 1969). The term *aeroallergen* has been used to refer to those allergenic materials which are transmitted through the air. Aeroallergens usually are hygroscopic solid particles having diameters between 1 and 80 μ. Most inhaled allergenic particles will exceed 1 to 2 μ in diam and have significant deposition in the upper air passages as a rule. Pollens are of sufficient size that removal by the upper air passages is virtually complete. Small spores may demonstrate considerable lung deposition. In the outdoor environment pollens, fungi, spores, insect emanations and algae particles are some important allergens. In the indoor environment, the most important aeroallergens are house dust and kapok, animal danders, fungi, spores, insect emanations, and seed proteins. Unlike some of the particulate matter previously discussed, the sensitizing particles may not act directly on the respiratory epithelium as their target tissue, but rather may react upon other target tissue, such as gastro-intestinal tract or mucous membranes of the respiratory tract. If the target tissue is the wall of the bronchial tract, the allergic reaction is referred to as asthma. Byssinosis is a respiratory disease of cotton textile workers resulting from an allergic response of the respiratory tract to a protein material. The target tissue is the wall of the bronchial tract, and bronchial constriction leads to airway resistance, and difficulty in breathing. Chronic exposure results in impaired pulmonary function and decreased ventilatory capacity. Lung diseases where the alveolar membrane is the target tissue are bagassosis and farmers' lung disease. In bagassosis the inciting agents are fungi which grow on bagasse, which is the residue of sugar cane following extraction of sugar. The farmers' lung group of diseases results from exposure to moldy hay and silage. Fever, cough, and difficulty in breathing follow within hours of exposure. Chills, lassitude, and headache are also commonly associated, as are granular or mottled areas of the lungs seen on the chest X-ray. Impairment of alveolar gas transfer and non-uniform gas distribution within the lungs is noted. There is minimal bronchial obstruction. The acute symptoms, though reversible, may become chronic if untreated or after long repeated exposure. Severe respiratory disability and even deaths associated with farmers' lung disease are prominently seen in agricultural areas of South America, Great Britain, and Northern Europe. The involved antigens are believed to be related to certain actinomycetes, especially Thermopolyspora and Micromonospora genuses. These organisms flourish when hay and other forage are heated during fermentation by fungi and bacteria. A syndrome quite similar in many respects to farmers' lung disease has also been found to follow exposure to moldy barley, oats, and corn.

C. EPISODES OF UNCERTAIN ETIOLOGY

Several episodes of acute air pollution, commonly found on a recurring basis, have been reported and the etiology of these episodes is still a puzzle. In New Orleans since 1955 there have been seasonal outbreaks of asthma. These usually occur in October and November and the characteristic pattern of the outbreaks is similar from year to year. Causative factors incriminated have been aeroallergens, particulate matter of undetermined type from burning refuse dumps, dust from flour mills, and others. 'Tokyo-Yokihama asthma' is another somewhat puzzling condition, the etiology of which has likewise remained uncertain. Persons afflicted respond favorably when removed from the geographic area, and the symptoms recur upon return to the area. In Minneapolis, in 1956, asthmatic attacks were found to be related to pollutants from neighboring storage and processing plants producing grain dust. The grain dust particles from Minneapolis caused reactions in subjects who had been affected in New Orleans and cross-reactions from the New Orleans dusts were noted in Minneapolis subjects.

D. INFECTIOUS DISEASES

Biological organisms which cause disease may be transmitted by spread through the air. Bacteria, viruses, and other living organisms which are pathogenic to man are most commonly transmitted by passage from one individual or source of infection to another. Aerosols are the most common mechanism for transmission of bacterial and viral infections. A classic example of airborne infection occurred on the U. S. S. Richard Bird, when 50% of the crew of this ship contracted tuberculosis from a single undetected case. It was found that the tuberculosis organisms were spread through the ship's ventilation system (Handler, 1970).

The atmospheric spread of infectious disease is determined by size and settling velocity of airborne particles, the variants of the organisms transferred in this manner, the susceptibility of the host, and the actual site of deposition of the particles. We have seen that particle size is of vital importance in providing access to the lower reaches of the respiratory tract. The small particle size of bacteria and viruses, together with droplet and droplet nucleae suspension capabilities of these particles, renders them capable of remaining in the atmosphere over long periods of time to be inhaled by susceptible individuals. Re-suspended particles from dust deposits on the floor or walls or ledges will be of sufficient size to be less of a factor in the spread of airborne infection than the direct spread of the finer particles dispersed from the respiratory tracts of infected individuals.

E. OCCUPATIONAL AND INDUSTRIAL EXPOSURES

Industrial exposures provide considerable evidence relating to the effects of acute high dose rate exposures, as well as chronic moderate dose rate exposures and chronic low dose rate exposures. However, it is unwise to depend upon industrial exposure data and to use industrially allowable exposure levels as community standards. In the first place, the workers are usually in a controllable occupational atmosphere which is confined in its exposure to a limited area. Intensity reduction is constantly practiced and with the increase in the capabilities of removing pollutants from the industrial environment, the environments are becoming less and less hazardous, for the most part. The exposed persons are usually in good health and if they have demonstrated some idiosyncracy to the particular pollutants involved they have been self-eliminated. That is, workers who are particularly sensitive to industrial dust from one source or another, will likely seek employment elsewhere if they have had excessive or serious symptoms from exposure to such industrial dust. The age of workers is of considerable importance. Those in hazardous areas are usually younger and healthier than the general population. It should be remembered that they work 40 hr a week in their occupational exposure. This leaves 128 other hr each week when there is no occupational exposure. Workers are usually exposed to a single pollutant, frequently in high concentrations.

Occupational exposures in individual cases and in studies of population groups must not be ignored. The potentiating or synergistic effects of exposures in the home, in the recreational environment, etc., must be further considered. Workers exposed to industrial hazards which promote the development of significant obstructive or other types of chronic lung disease are in a more hazardous situation if they must also live in an urban environment which is heavily contaminated with respiratory irritants. Indeed, workers, the backbone of our economic system, may be placed in a double jeopardy situation. That is, they may be exposed to the ambient level of pollution in their living area and then they may be exposed to added heavy doses of sulfur dioxide, carbon monoxide, lead, smoke, dust, fumes, and other contaminants in their work environment. Double jeopardy may even be expanded to triple jeopardy if one adds the additional hazard to health from the personal pollution of cigarette smoking. Hence, they may fit into those groups of our population with what we consider to be increased susceptibility to the effects of air pollution, such as premature infants, newborns, the elderly, the chronically ill with cardiorespiratory disease, etc.

Aerial spraying of pesticides and agricultural chemicals is thought of as an occupational exposure to toxic chemicals. This occupational hazard may extend to the general environment. Pesticides are used on a world-wide scale for disease control and crop production enhancement. Malaria, yellow fever, plague and dengue are only some of the more important diseases which have been controlled or even completely eradicated in some areas of the world by use of chemicals which eliminate insects and rodent vectors of disease. There are thousands of pesticides used to increase and optimize crop production. Exposure may occur both in occupational and non-occupational situations. Occupational exposures can occur in those persons engaged in the development, manufacture, distribution, and sale of chemicals if they are not subjected to proper protective procedures. The application of pesticides in agriculture can lead to occupational exposures in crop dusting pilots, crop workers, farmers, nursery workers, and to the general population in the use of pesticides in gardening and yard maintenance. Additionally, fumigators and exterminators may sustain occupational exposures to dangerous pesticides.

6. Effects of Known Toxic Agents in Air Pollution

The discussion to follow will be subdivided into an analysis of the effects of known toxic agents which are in the gaseous state, and the known toxic agents in the particulate state. The particulate discussion will include aerosols and fumes. No effort will be made to present an encyclopedic listing of observed or measured relationships between specific levels of pollutants and signs and symptoms of disease. The interested reader may refer to detailed treatment of these relationships in the literature, particularly in the U. S. Department of Health, Education, and Welfare publications dealing with the specific pollutants. The purpose here is to present some representative findings only to illustrate magnitudes of concentrations which are of concern rather than to list finite figures.

A. GASEOUS TOXINS

Ozone and Photochemical Oxidants – Data on ozone effects on humans come primarily from occupational exposures and human experimentation. Effects are not noted at concentrations up to 390 μg/m^3 (0.2 ppm). The threshold of nasal and throat irritation appears to be at about 590 μg/m^3 (0.3 ppm). Exposure to a concentration of 980 μg/m^3 (0.5 ppm) for 3 hr a day, 6 days a week has resulted in a decrease in the one second forced expiratory volume after an exposure time of 8 weeks. Exposure to concentrations from 1180 to 1960 μg/m^3 (0.6 to 1.0 ppm) for 1 to 2 hr may possibly impair pulmonary function by the mechanism of increased pulmonary resistance, decreased carbon monoxide diffusing capacity, decreased total capacity, and decreased forced expiratory volume. Exposure to concentrations from 1960 to 5900 μg/m^3 (1.0 to 3.0 ppm) for ten to 30 min is intolerable to many subjects. Exposure to a concentration of 17 600 μg/m^3 (9.0 ppm) will produce severe illness. Experimental exposures of humans in a wide range of concentrations have demonstrated progressive changes in pulmonary ventilation impairment in a dose-response relationship.

The clinically recognized symptoms from inhaling ozone begin with detection of odor, and continue through dryness of mucous membranes of the upper respiratory tract, changes in visual acuity, and functional derangements of the lung, pulmonary congestion, and edema (USDHEW, 1970). The site of action of ozone appears to be at the level of the cell membrane, with a breakdown of unsaturated fatty acid in the cell membrane (Goldstein *et al.*, 1970; Balchum *et al.*, 1971). Experimental data are also available to show that ozone and nitrogen dioxide increase the susceptibility of laboratory animals to respiratory infection when the animals are challenged by inhalation of pathogenic organisms. The damage produced appears to be temporary, at least with repeated short exposures repeatedly (Stokinger, 1957; Stokinger *et al.*, 1957; Stokinger and Coffin, 1968).

Ozone is an important constituent of photochemical smog. It appears to be the major component of photochemical oxidant. Other free radical oxygen forms, peroxy-acetyl nitrates (PAN homologs), and certain oxides of nitrogen comprise the remainder of photochemical oxidant. Total oxidant levels appear to be very important in the assessment of effects on humans. Hence, although ozone is the major component of photochemical oxidant, it appears that a total oxidant level somewhat below that which would be measured for ozone alone causes human effects.

Nitrogen Oxides – There are several commonly occurring oxides of nitrogen. For purposes of our discussion, nitrogen dioxide (NO_2) will be considered the prototype, since the other oxides of nitrogen react in air to produce NO_2. Nitrogen dioxide produces nose and eye irritation and with increasing concentration and times of exposure, obliterative bronchiolitis, pneumonitis, and even death. The odor threshold for NO_2 appears to be approximately 1 to 3 parts per million. Nitrogen dioxide appears to be capable of producing a variety of clinical reactions and with different clinical syndromes resulting from different degrees of exposure. Severe exposure (concentrations of 500 ppm or greater for short periods of time, a few minutes to an hour) resulted in acute pulmonary edema with bronchial pneumonia and death in a few days in animals. In lesser ranges, such as 150 to 200 parts per million, fibrous and obliterative bronchiolitis developed and was fatal in three to five weeks. Fifty to 100 parts per million can bring about the development of bronchiolitis with focal pneumonitis lasting from 6 to 8 weeks, usually with spontaneous recovery. Reactive hyperplasia of both the terminal bronchiolar and alveolar epithelium has been well proven in definitive experimental studies. A relatively low level (15 ppm) of NO_2 has been shown to produce a measurable increase in the hyperplasia of pneumocytes. The occurrence of Type 2 pneumocyte hyperplasia in response to nitrogen dioxide inhalation has been quantitatively demonstrated by electron microscopy. Thickening of the blood-gas barrier through replacement of the thin type cells of the alveolar wall with cuboidal or columnar type cells (Type 2 pneumocytes) has been implicated in the adverse effect of 10 ppm NO_2 (Yuen and Sherwin, 1971).

Lowry and Schuman (1956) described 'silo fillers' disease', a syndrome caused by NO_2. Nitrogen dioxide is formed from fresh green silage when confined in the silo. Exposure to the gases in silos has been followed almost immediately by symptoms of malaise, cough, shortness of breath, chest pain, chills, fever, nausea, and vomiting. The signs may last for a few days to a few weeks. There may be complete resolution of these symptoms or there may be a progression of the condition to a severe and potentially fatal pulmonary insufficiency. Findings of obstructive pulmonary function are present and the pathologic findings are those of a fibrous obliteration of the bronchioles.

Oxides of Sulfur – A number of oxides of sulfur exist; sulfur trioxide, its hydrate, sulfuric acid as a mist, and as a sulfate are of concern. We will use sulfur dioxide (SO_2) as the model for discussion of the effects of oxides of sulfur. Sulfur dioxide in concentrations below 25 parts per million appears to have irritant effects confined to the upper respiratory tract; however, if particulate adsorption takes place with the formation of aerosols, the lower portion of the respiratory tract may be involved. Virtually complete absorption of the gas by the upper airways raises considerable question as to how to explain the increase in pulmonary flow resistance. It has been postulated that there may be a reflex change in bronchomotor tone (Nadel *et al.*, 1965).

The lowest concentration known to elicit a human response is approximately 0.2 ppm of SO_2. This is a level reported by Soviet scientists and was considered to be a threshold for inducing conditioned reflexes in the brain cortex. For taste, the threshold is usually considered to be 0.3 ppm which is characteristically lower than the odor threshold of about 0.5 ppm. Concentrations above 1 ppm of pure SO_2 are probably necessary before serious or significant effects could be expected on the health of unimpaired individuals. Industrial exposures to men of levels up to 36 ppm have resulted in chronic effects of high incidence of nasal pharyngitis, cough, expectoration, shortness of breath, and other signs of respiratory distress. However, these levels are much higher than would be foreseeable for ambient air pollution (USDHEW, 1969).

Frank (1964) reported a number of studies performed over several years to investigate the physiological response of the lungs to SO_2. The human subjects were volunteers who were felt to be free of underlying pulmonary disease. The range of SO_2 used was that which is generally encountered in urban air pollution. Using pulmonary flow resistance (PFR) as a single parameter, one subject was found to demonstrate an increase in PFR in response to 1 to 2 ppm of SO_2. In response to 5 ppm of SO_2 for 10 min there was an average increase of PFR by 30% above control levels. At 13 ppm there was a rise in the group to 72% of those control levels. This dose-response relationship is practically linear. To demonstrate the importance of absorption of SO_2 in the upper respiratory tract comparisons were made by administering SO_2 by mouth and by nose. The PFR increased significantly in 9 of the 12 experiments in which SO_2 was administered by mouth, but only increased in 3 of 12 exposures to the same levels of gas by nose. The net uptake by the airways was calculated to be 84%. The techniques also demonstrated that not only do the upper airways absorb virtually all of the SO_2 administered but that there appeared to be some release of the gas also during expiration.

Potentiation of SO_2 effect is known to be influenced by the presence of particulate matter, with a three- to four-fold potentiation of the irritant response to SO_2 in the presence of particulate matter which is capable of oxidizing SO_2 to sulfuric acid especially. Aerosols and soluble salts of ferrous iron, magnesium, and vanadium have been observed to produce the potentiation although the concentrations used were greater than any levels of these metals usually reported in urban air (Amdur, 1969).

Carbon Monoxide – Carbon monoxide (CO) is a colorless, non-irritating gas which is generated by incomplete combustion. It is ubiquitous in nature and is a major component of motor vehicle exhaust. No other gaseous pollutant with as much toxic potential exists at such high concentrations in urban atmosphere. The current levels of urban air pollution of this compound have been reported at 10 to 30 ppm, with peak concentrations reaching 75 to 85 ppm during rush hour traffic in poorly ventilated city areas. Carbon monoxide is ten times more toxic than SO_2. Sustained atmospheric levels of CO of about 10 ppm will produce a carboxyhemoglobin level of about 2%. Behavioral performance has been shown to be degraded by 2% carboxyhemoglobin blood levels (NAS-NAE, 1969).

Carbon monoxide impairs oxygen transport of the blood. The affinity of hemoglobin is 210 times greater for CO than for oxygen and carboxyhemoglobin is known to interfere with the release of oxygen carried by the hemoglobin molecule. The mechanism of action is the impairment of transport of oxygen at the tissue level. Hence, at high altitudes and other situations where oxygen tension is low, there may be a detrimental effect. Persons with impairment of the circulation or severe anemias contribute to special sensitivity groups. Increasing concentrations of CO induce signs and symptoms of headache, dizziness, lassitude, ringing in the ears, nausea, vomiting, palpitations, pressure in the chest, difficulty in breathing, apathy, muscular weakness, collapse, unconsciousness, and finally death. The 'background level' of carboxyhemoglobin is about 0.4% in the blood. This level is the result of the balance between endogenous CO production, endogenous CO metabolism, and elimination of CO through the lungs. Smoking a pack of cigarettes a day will produce carboxyhemoglobin levels of about 5%. It is likely that these levels alone might be injurious to health, especially in the presence of conditions where there is less than the normal amount of oxygen available to the tissues, such as coronary vascular disease, peripheral vascular disease, emphysema and asthma. In pregnancy, the tissue oxygen tension in the fetus is less than that of the maternal oxygen tension owing to the presence of carboxyhemoglobin in the fetal blood. The placental blood of a mother who has a high carboxyhemoglobin level in her blood would thus lower the oxygen supply available to the fetus. Mothers who smoke have babies with lower average birth weights than non-smoking mothers.

Ayres *et al.* (1970) reported studies which were performed before and after the administration of CO for varying time periods to patients with coronary artery disease and patients with other cardiopulmonary disorders. Coronary blood oxygen tension decreased in all but one of the patients. Patients with coronary artery disease are particularly vulnerable to the stresses of induced hypoxia because the rigid vascular bed may not adequately dilate to allow more blood flow to hypoxic areas. Myocardial muscle surrounding a diseased vascular supply can become critically hypoxic within a small area of the heart even with low levels of carboxyhemoglobin. A single focus of hypoxia is felt to be responsible, in many cases, for the initiation of fatal episodes of ventricular fibrillation. The influence of cigarette smoking in adding additional hazard to coronary heart disease risks from CO exposure is obvious. Numerous investigators have shown that non-smokers in comparable groups run consistently lower median carboxyhemoglobin percentages than smokers by a magnitude of about two to three times. Webster (1970) reported recently that squirrel monkies fed cholesterol and exposed to CO developed a greater degree of coronary artery stenosis than controls fed cholesterol without CO exposure.

An association between areas of Los Angeles having higher CO levels in the ambient atmosphere and higher case fatality rates from myocardial infarction has been demonstrated by several studies (Cohen *et al.*, 1968; Goldsmith and Landaw, 1968). Other evidence points to the possibility that daily average CO levels in excess of 10 ppm may be related to increased mortality in those patients who have been hospitalized with myocardial infarction.

Speculative evidence has been advanced that there may be a relationship between automobile accidents and high blood levels of carbon monoxide. However, further study is indicated, since some of the data used to reach this type of hypothesis has indicated that studies were not controlled for cigarette smoking.

Hydrocarbons in Gas Phase – Hydrocarbons with a carbon number greater than 12 are generally not encountered in the atmosphere in concentrations of sufficient intensity to be of concern. Natural sources of hydrocarbon emissions are usually from living organisms. Hydrocarbons promote the formation of photochemical smog, and are felt to contribute to the development of eye irritation and other eye manifestations. Formaldehyde, the aldehydes of ketones, and various oxidants, including peroxyacetyl nitrate, are formed from hydrocarbons. No effects have been reported at levels below 500 ppm for the aliphatic and alicyclic hydrocarbons; hence, they may be generally considered to be of little biologic significance.

There is little evidence to indicate direct health effects of the gaseous hydrocarbons present in the ambient air. The effects attributed to photochemical smog must be considered indirectly related to the ambient levels of the hydrocarbons, since the aromatic hydrocarbons, which are biochemically and biologically active, have properties which cause irritation of the mucous membranes and in sufficient concentrations may cause systemic injury.

Hydrocarbons: Miscellaneous Irritants – Ketones and ketenes are known toxins. The formation of ketones as the product of chain reactions of free radicals in photochemical smog is known to occur, but there have been insufficient levels measured to have any known effect. Ketenes have not been identified in photochemical smog but can be shown to result from the action of ozone on olefins in the laboratory. Ketene is an extremely lethal chemical substance and has high irritant effect. Little attention has been devoted until recently to the organic sulfur compounds such as the dimethylsulfide, methylmercaptan, and the thiols. Sketchy information to date leaves little doubt that these substances can be highly toxic; however, no known concentrations have been detected in air pollution to cause concern.

B. PARTICULATES

For purposes of brevity, aerosols and fumes will be combined under this general category of particulates.

Undifferentiated Particulate Matter – Combinations of particulates dispersed in the atmosphere consist of a large variety of substances, among which are fluorides, beryllium, lead, asbestos, fly ash, and other organic and inorganic materials. Undifferentiated particulate matter as discussed here, refers to any particulate matter dispersed in the air, solid or liquid, in which the individual particles are smaller than 500 μ in diam and larger than molecules. Some of the specific known toxic agents, such as beryllium, lead, etc., will be discussed in more detail.

The effects of particulates on the economic and aesthetic values of the community are related to visibility, soiling, corrosion, and other effects. Adverse health effects appear to be related to injury to the surfaces of the respiratory system. The injury may be permanent or reversible. Particulate matter may produce direct injury upon the surface of the respiratory tract, or the injury may extend beyond the surface of the tract and may act in conjunction with gases altering their sites of action. A very important element in the toxicology of particulate matters is their fate in the body (deposition, disposition, and retention). Synergistic effects of particulates with other pollutants are of considerable importance.

Relationships for short term air pollution episodes are demonstrable and indicate changes in death rates coincident with high levels of pollutants. Excess deaths and increase in morbidity have been observed in London and New York with smoke levels above 75 μg/m^3 or smoke shade index above 5 or 6 cohs, respectively. Sulfur oxide levels were high in both cases.

In England, exacerbation of acute bronchitis conditions has been associated with daily averages of smoke above a 300 μg/m^3 to 400 μg/m^3 level. Increased absences due to illness were noted in England when smoke levels exceeded 200 μg/m^3. Other British studies have shown an increase in selective respiratory illness in children to be associated with annual mean smoke levels above 120 μg/m^3. In the United States increased death rates from selected causes in males and females to 69 yr of age were felt to be related to annual geometric means of 100 μg/m^3 and over of particulates. Corroborating evidence has been supplied by a study in Nashville, Tennessee. In both of these examples, sulfur oxide pollution was present. Another study in London showed that there was a decrease in mean sputum volume as the smoke pollution (yearly mean) decreased from 140 μg/m^3 to approximately half that level. The investigators noted however that there had been changes in the tar composition of cigarettes during the same period, an uncontrolled variable that could affect the findings (USDHEW, 1969). Winkelstein and Kantor (1969) studied mortality rates for gastric cancer in white men and women, age 50 to 69. The rates were found to be almost twice as high in areas of high suspended particulate pollution as in areas of low pollution in a northeastern United States industrial area. The effects of economic status and ethnic distribution of the population appeared to be independent of the association between mortality rate and air pollution. Winkelstein and Gay (1971) found that age- and sex-specific death rates among whites for cirrhosis of the liver for the pericensal period (1959–1961) showed a graded positive association with suspended particulate air pollution. The association was found to be strongest in the lowest economic groupings. They postulated that their findings indicated that urban particulate air pollution may contain certain toxic agents capable of acting directly or synergistically to cause cirrhosis of the liver. The synergistic relationship of particulate matter which is capable of oxidizing SO_2 to sulfuric acid has been previously discussed. On the basis of the toxicologic evidence and the literature of atmospheric chemistry, the levels of both particulates and oxides of sulfur should be determined in an associative manner, and not on the basis of the individual elements.

Fly ash is the generic term applied to the material emitted from the stacks or chimneys of furnaces which burn fossil fuel. A large portion of airborne particulate matter is made up of fly ash. The composition of fly ash will vary depending upon the type of fuel and method of firing used. Silica, alumina, iron oxide and carbon are the usual substances found in fly ash. The association of fly ash and health effects is vague, except where decreased visibility affects the well-being of individuals.

Carbon – Coal dust is a complex substance consisting not only of the carbon which rises from the disintegration of coal but dusts containing silica and dusts of various rocks in and above the coal seam. *Anthracosis* is a term referring to a blackish pigmentation of the lungs caused by deposition of coal particles. There is usually no associated pathologic change in the lungs and the condition is frequently observed in persons who live in areas of heavy industrial pollution. The term *anthracosilicosis* refers to a condition caused by exposure to coal dust and rock dust containing free silica. The condition is accompanied by extensive fibrosis and closely resembles the changes of silicosis, but carbonaceous material is also present.

Coal workers' pneumoconiosis is a separate and distinct condition from anthracosis or anthracosilicosis. The condition is caused by coal dust with little or no pathology attributed to silica. The delineation and identification of this condition, at least in the United States, have been quite slow in developing. The British recognized the condition in the late 19th Century. The confusion of the findings of coal workers' pneumoconiosis with combined and distinct cases of silicosis left the recognition of the disease as a separate entity somewhat questionable. Since the early 1950's there has been a general recognition of the presence of this condition as a disease entity and an increasing number of cases are being reported in the literature. The diagnosis of the disease depends upon characteristic chest X-ray findings and an occupational coal dust exposure. There are attacks of difficulty in breathing, frequent cough, expectoration of inky black sputum, severe breathlessness on exertion, and finally the condition may become disabling, with the development of a condition known as focal emphysema. Extensive fibrosis of the lung is frequently complicated by the superimposition of tuberculosis which leads to a rapid deterioration and death. The combined pathological findings of coal workers' pneumoconiosis and tuberculosis lead to a condition called progressive massive fibrosis (PMF). The cause of simple coal workers' pneumoconiosis has been attributed to an overwhelming accumulation of fine particles of coal dust with the accompanying mechanical overloading of the lungs (Kerr, 1956, 1970). Coal workers' pneumoconiosis serves as an example of a disease which develops upon extreme and heavy exposure to a substance which, in our average environmental exposure, is of little or no consequence; however, it is not an insignificant disease. Kerr (1969) has conservatively estimated that there are approximately 50 000 active and former coal miners disabled by the disease. That the condition is preventable is indeed a true example of 'ill-informed complacency' of occupational health and preventive medicine workers.

There is no supportable evidence that exposure to coal dust, *per se*, in the small amounts found in the usual urban atmosphere is harmful as a distinct toxic entity. Carbon alone, though it is generally felt to be innocuous, can, in overwhelming doses, be responsible for slowing pulmonary clearance rates and producing a longer action of any other agents which may be present in the more vulnerable portions of the lungs. It also conceivably can serve as a means of increasing or enhancing the penetration of other substances through the alveolar membranes.

Asbestos – Asbestosis, the condition caused by pathologic changes in the lungs secondary to toxicity of asbestos, has been known as an industrial hazard for many years. The condition develops after a long latent period and is felt to require a prolonged period of exposure (as much as 10 to 20 yr) to result in the progressive diffuse fibrosis that characterizes the disease. The clinical picture for asbestosis is never typical. The insidious onset and slow progression of the disease with possible super-imposition of other lung diseases, make the condition difficult to describe and make its course quite variable. Recently, there has been an accumulation of evidence which implicates chrysotile asbestos fibers in the lungs with the development of diffuse mesotheliomas of the pleura or peritoneum. In South Africa over a 4 yr period 47 cases of mesothelioma were discovered. Of these, some 45 had had asbestos contact. No mesotheliomas were reported in the rest of South Africa. There was also no development of mesothelioma reported in some 10 000 silicosis autopsies. In London, 76 cases of mesothelioma were investigated, all of which seemed to be closely related to living in the proximity of an asbestos plant (Selikoff, 1970). Like asbestosis, meso-thelioma also has a very long latent period. Increasing exposure of the general population to asbestos may thus be presenting a hazard which we will not see and be unable to measure for some time. Construction asbestos spraying began in 1959, hence it may be 1979 or 1989 or even the year 2000 before we begin to see an increase in mesotheliomas if the environmental exposure is an important factor in producing this condition. Langer *et al.* (1971) subjected 28 consecutive New York City autopsy cases to electron microscope examination of representative small lung samples. In 24 of the 28 cases chrysotile asbestos fibers were unequivocally demonstrated. Similar findings have been made in London. There is no reason to believe that this finding would not hold for other urban areas. It is of interest to note that New York City has an ordinance which prohibits the open spraying of asbestos in construction work, so as to protect the general population. The source of air pollution from asbestos fibers results from not only exposure in industry but as indirect occupational exposure in construction workers, such as those working with asbestos cement, and as a direct exposure in construction workers spraying asbestos. Serpentine outcroppings of asbestos in its natural state can also be a source of exposure. Weathering of asbestos shingles may provide exposure to the fibers.

Other Dusts – For the most part pneumoconioses are seen only in heavy industrial exposures, and hence are of only passing interest in the consideration of air pollutants and their effects on the general population. Some of the more commonly seen pneumoconioses include silicosis, coal workers' pneumoconiosis, asbestosis, diatomite pneumoconiosis, shavers' disease, byssinosis, bagassosis, and farmers' lung. We have discussed some of these earlier. Some of the rarer conditions include pneumoconioses from mica, cement, gypsum, fluorspar, sulfa, jute, and grain. These conditions will not be discussed here and the reader is referred to appropriate textbooks on occupational diseases for further information.

Beryllium – Beryllium metal and its inorganic compounds are of interest for two reasons. First, beryllium exposure was found to cause a progressive and very serious, frequently fatal, lung disease in beryllium workers, and then was later demonstrated to be the causative agent in similar conditions in persons exposed downwind from beryllium plants and industrial operations where beryllium was used, such as the production of fluorescent lighting bulbs. Persons residing in homes with beryllium workers were also affected. Second, beryllium is widely used in production of industrial materials and in nuclear reactor work. There is an increasing interest in the use of beryllium and its compounds in industrial, military, and space operations. These possible expansions in the use of beryllium may create an increase in the likelihood of exposure. Beryllium can cause local effects on the skin by producing a dermatitis felt to be primarily irritant in nature. Granulomatous lesions or ulcers can be produced from this contact. Irritation of the conjunctiva of the eye and irritation of the cornea can follow contact with beryllium salts. Nasal irritation with rhinitis and pharyngitis may also result from exposure. The primary route of entry however is inhalation of the fume or dust which produces a systemic disease of either acute or chronic nature. In the acute disease there is the development of progressive signs of shortness of breath, cough, loss of appetite, and weight loss. Pulmonary involvement can lead to symptoms of pneumonitis and there may be rapid progression of symptomatology. Fatal termination can result from cardiopulmonary involvement but recovery is complete within a period of about 6 months in many cases. Chronic beryllium poisoning can result in granulomatous lesions of the skin, liver, kidneys, spleen, and lymph nodes, but the most important condition, again, is the development of pulmonary lesions with a chemical pneumonitis or bronchial alveolitis. The symptoms are usually delayed in onset in the chronic disease but are more persistent in nature and the case fatality rate may reach as high as 35%.

There is increasing interest in beryllium disease from the standpoint of the possibility of the long term effects influencing the development of cancers. No firm data are available at this point in time to implicate beryllium as a causative factor in cancer.

However, there has been demonstrated recently an indication that prior chemical respiratory illness may influence subsequent development of lung cancer among beryllium workers (Mancuso, 1970). Air samples have demonstrated the presence in nanogram amounts in the urban atmosphere of a number of cities. Although the concentrations were in very small amounts a continuous inhalation of the maximum air levels found could result in pulmonary retention of approximately $\frac{1}{2}$ mg in 30 yr, an amount which could be toxic (Schroeder *et al.*, 1970).

Lead – The total body burden of lead is derived from food, water, and inspired air. Compared with the daily lead intake from food and water variously estimated to range between 0.1 and 0.4 mg per day, that taken in the inspired air ranges from about 0.01 to about 0.10 mg per day (USDHEW, 1970). Only 5 to 10% of the lead taken in by consumption of food and water may be absorbed in the body but between 30 and 50% of inspired lead is absorbed by the lungs. Those forms of lead which are of interest from a viewpoint of air pollution are the metal itself, the various oxides of lead and several mixed salts derived from the burning of gasoline additives, tetraethyl lead and tetramethyl lead. The signs and symptoms of lead poisoning can include abdominal pain and tenderness, constipation, headache, weakness, muscular aches and cramps, vomiting and anemia. There is frequently a characteristic lead line on the gum margin. Paralyses and encephalopathy can result from extreme lead poisoning. Children are particularly susceptible to lead poisoning, since they are known to chew on furniture and other painted surfaces which may be painted with leaded paint. A very large number of potential industrial exposures are possible.

Chow and Earl (1970) measured the concentrations of atmospheric lead in the area around San Diego, California, and found that the lead aerosol concentration is increasing at a rate of about 5% per year. The isotopic composition of the aerosols is similar to the additives of gasoline which are, of course, the largest contributors to atmospheric lead. Lead blood levels are higher in urban areas than in rural areas of the United States. There appears to be a fairly close relationship between the atmospheric exposure levels and the blood level. In Los Angeles it was found that the average blood levels in the population living near a freeway were higher than those in persons living upwind from the city and near the beach. The body burden of lead is generally related to the blood level, and most authorities concur that increase in the body burden of lead is undesirable. There would seem to be very sound reasoning behind recent Government and petroleum industry moves to reduce the lead levels in gasoline.

Other Metals – Mercury and its compounds can produce abdominal pain, vomiting, diarrhea, pneumonia, renal damage, and circulatory and respiratory failure in acute severe exposures. Chronic exposures can lead to conditions manifested by signs of gingivitis, tremor, and emotional instability. These are usually accompanied by headaches, digestive disturbances, renal damage, and restriction of visual fields, among other symptomatology. Inhalation of the fumes or percutaneous absorption of the metal and organic compounds of mercury is the route of entry. The problem of mercury is primarily an occupational condition. Mercury is not held to be a significant air pollutant, inasmuch as it is usually available in only small amounts and does not become a general atmospheric pollutant.

Rarely or infrequently encountered metals which have toxic properties are antimony, bismuth, and tin. Beryllium can be included in this group. Antimony is a relatively toxic metal but it is not known to be present in air in any dangerous or hazardous amount. Bismuth has relatively low toxicity but can cause renal and liver damage in large doses. Tin has a low toxicity and it is doubtful that tin in the amounts common in the atmosphere could conceivably cause any hazard. Nickel will produce cancer in both animals and man when inhaled as nickel carbonyl, hence it does represent a true hazard to human health. An important source of nickel is the exhaust from burning diesel fuel. It can also come from burning coal as well as from additives to gasoline, incineration, and emissions from oil refineries. Hence, nickel may be a hazard to populations and should be controlled. Cadmium is found in the air and in milk. Cadmium has been postulated as a cause of hypertension and it is absorbed from the lungs to be deposited in the kidney, liver, and pancreas (Schroeder *et al.*, 1970). Little evidence has accumulated to date to support this hypothesis, so further data are needed.

7. Effects of Combined Air Pollutants

A. ODOR

The production of odors by air pollutants has been frequently viewed as a manifestation of nuisance value only. The detection of a pollutant by its characteristic odor does not necessarily indicate that the concentration of the pollutant is sufficient to cause health effects. However, in our definition of health, the nuisance effect of odor is a legitimate consideration as a health effect. Detection of odor is not a reliable technique for identification of health hazards. For example, the nose rapidly becomes tolerant to low levels of hydrogen sulfide and other substances. Hence, the concentration of the gases may increase without odor detection, if an individual has been subjected to a slowly rising concentration. There are gaseous products, fumes, and other air pollutants which have no odors. Carbon monoxide, unquestionably a serious toxic agent, is odorless. A number of atmospheric contaminants such as hydrogen sulfide, the mercaptans, ozone, the organic peroxides and many other organic compounds produce unpleasant odors. These may be very disagreeable and are often the chief cause of annoyance among persons residing in communities where such contaminants are discharged into the atmosphere. Offensive odors may lead to a loss of appetite or nausea, and may be a significant factor in causing residents to move from an affected area.

Some substances are notoriously identifiable by their odor. The rotten egg smell of hydrogen sulfide is an outstanding example. The odor detection point of this chemical is very close to the toxic level. Ozone, on the other hand, has an odor detection level lower than that of any detectable physical or medical effect. The level for odor detection of ozone is between 0.02 and 0.05 parts per million for the general population. Certain sensitive individuals do report irritation of the nose and throat at the 0.05 ppm level, but for the most part irritation effects occur at 0.1 ppm and above (Stokinger and Coffin, 1968). Sulfur dioxide has a threshold for taste commonly considered to be about 0.3 ppm which is characteristically lower than the odor threshold which is about 0.5 ppm. Measurable effects in the sensitivity of dark adaptation of the eye are notable at a level slightly above 0.3 ppm. Hence, it may be that there are measurable physical or medical effects very close to the odor threshold of SO_2. Odor is a noticeable effect of NO_2. Pure NO_2 has a range of odor threshold reported to be between 1 and 3 ppm. Serious medical and physical effects are known for NO_2 in the higher concentrations, but there has been no solid evidence to in- criminate NO_2 effect at the odor threshold.

B. VISIBILITY

Deterioration of visibility is an obvious manifestation of the effects of air pollution. Visibility may be affected over a vast expanse such as a city or valley, or it may be affected in highly localized areas such as one adjoining a factory. Most air quality standards reflect the visibility criteria in the particulate standards. For example, in the California State Department of Health standards for ambient air quality, 'adverse' levels are considered to have been reached when the particulate concentration is sufficient to reduce visibility to less than 3 miles when relative humidity is less than 70% (Stokinger and Coffin, 1968). Visibility is dependent both upon the nature of whatever particulate matter may be present in the atmosphere and also upon the volume of air with which the material is mixed. Aerosols are considered to be the single most important factor in determining the loss of visibility in polluted atmos- pheres. Sky color effects, especially the brown discoloration resulting from high con- centrations of NO_2, are also considered to be undesirable. As with the discomfort produced by odors, the impairment of visibility is oftentimes sufficient to lead individuals to change their residence or place of employment. Visibility may play an extremely important and critical role in its relation to automobile and aircraft accidents if the impairment of visibility is sufficient to decrease safety of operation of automobiles, as on freeways, or aircraft taking off or landing at airports in urban areas, or in the increased likelihood of mid-air collisions in crowded airways over cities.

C. EYE IRRITATION

One of the hallmarks of photochemical smog typical of the Los Angeles Basin is the presence of compounds apparently due to oxidation. Besides causing damage to field crops, ornamental plants and trees, reducing visibility and causing loss of vitality of an individual exercising on a smoggy day, they also cause eye irritation. Eye irritation is a known and measurable phenomenon. Interestingly, those components of photo- chemical oxidants which are specifically responsible for eye irritation have so far eluded exact delineation. The effective eye irritants are thought to be the products of photochemical reactions, but a direct cause-effect relationship has not been demonstrated and the precursors of eye irritants appear to be organic compounds in combination with oxides of nitrogen, the most potent being aromatic hydrocarbons. The chemical identities of irritants in synthetic systems under laboratory conditions are formaldehyde, peroxybcnzoyl nitrate (PBzN), paroxyacetyl acetate (PAN), and acrolein.

D. ACUTE EPISODES

The problem of air pollution has commanded the attention of health authorities and other scientists for only a few decades. Before health workers became involved atmospheric pollution was handled as an economic problem. Most industries discharged their waste products into the air without regard for the effect on surrounding communities. Smoke abatement ordinances, when passed, were usually ineffective and only rarely enforced. Stimulus to the interest of health authorities in air pollution came largely through the dramatic and tragic acute episodes where obvious discernable mortality was found. The Meuse Valley episode, the Donora, Pennsylvania episode, the London-New York, and world-wide episodes are of particular significance to this discussion.

In early December 1930, a thermal inversion layer confined the pollutants of a number of industrial plants to a small valley of the Meuse River in Belgium. On the third day of the tragedy a number of people were stricken and before the week was out sixty had died. Illness affected persons of all ages. Those with known diseases of the heart and lungs and older individuals were primary victims (Firket, 1931). Considerable dispute over which chemical substances may have been responsible for the illness and fatalities continues to this day. Sulfur dioxides and fluorides have been implicated. Most likely a combination of several pollutants was associated with this disaster.

In 1946 the Los Angeles smog situation became acute, attracting the attention of health authorities in the United States. In October 1948, the attention of the entire country was focused on an episode which occurred in Donora, Pennsylvania. Stagnant meteorological conditions and low temperature in the valley of the Monongahela River led to a smog which lasted for $4\frac{1}{2}$ days. A marked increase in illness began on the second day and 17 deaths occurred on the fourth day. Altogether, twenty persons from ages 52 to 84 yr died and many others became severely ill. Pre-existing disease of the cardio respiratory system was present in the majority of the fatal cases. The symptoms of the illness included irritation of the respiratory tract and other mucous membranes. Autopsies showed edema, hemorrhage, purulent bronchitis, and bronchiolitis (Roueche, 1947; Schrenk, 1949). The exact cause of the illnesses was not apparent, but the evidence of respiratory tract irritation was clear. Studies appear to indicate that no single substance existed in sufficiently high concentration to be considered *the* responsible factor, but the more likely combination of a number of irritant gases and fumes caused the illness and deaths.

In London, England, during the period December 5 to 9, 1952, a peculiarly dense fog occurred. Excess illness became noticeable on the third and fourth days of the fog. Examination of the mortality during and following this period in London in comparison with previous years indicated that approximately 4000 deaths could be attributed to the fog. The increase in deaths paralleled the increase in concentration of SO_2 and smoke in the atmosphere. The deaths decreased subsequent to the fall in concentration of these substances. A number of domestic animals were also severely affected (Ministry of Health, 1954). Once again, the specific causes and agents were not clear. Comparable, but less serious episodes have occurred in London on several other occasions. A similar incident probably occurred in New York in November of 1953.

A striking world-wide episode occurred in November and December 1962. In the eastern United States during the period of November 27 through December 5, there were increased pollution levels from Cincinnati to the east coast. Respiratory symptoms were noted, especially among groups of older persons. No specific delineation was made pertaining to the number of excess deaths. In London, England, from December 5 to 7, the SO_2 levels increased and there were 700 excess deaths attributed to the episode with an increase in morbidity. The sulfur oxide level increased considerably in Rotterdam from December 1 through December 7 and there was an increase in mortality and hospital admissions, especially in older persons with respiratory conditions. Sulfur dioxide levels also rose in Hamburg, Germany between December 3 and December 7. Small increases in mortality were thought to have occurred in that city. In Osaka, Japan, during the period of December 7 to 10 the mortality study there demonstrated sixty excess deaths. The world-wide episode was best demonstrated in the areas delineated above. It should be understood that there were areas in the world during this period where no serious pollution was reported.

There is a common thread of evidence running through all of these episodes. The evidence is somewhat less specific than would be desirable, but there is a very clear association between excess mortality and morbidity and demonstrated air pollution events. Unfortunately, the other common thread that runs through all of these episodes is the basic lack of good quantitative data. There were no effective programs for monitoring daily levels of air pollution and for correlating these daily levels with morbidity and mortality data. Many such episodes have occurred in places other than those where adequate documentation has been accomplished and in many communities they may be taking place now without adequate documentation. Unless suitable investigations were made, the occurrence of acute episodes of air pollution associated with increased death and illness will not be known to have occurred. With the current widespread availability of some air pollution measurements and of weather forecasting and reporting, including a system of weather alerting for expected periods of stagnation, one phase of the program is relatively easily handled. The second part, that of morbidity and mortality incidence, is available but is less readily obtained. All large communities should be alerted to the need for determining whether or not such episodes have occurred or are occurring in their own areas. Studies connecting acute episodes with morbidity and mortality at the time the episodes occur will be immensely more valuable than retrospective studies in delineating cause and effect relationships.

In addition to these and other episodes of general air pollution, acute exposures to specific materials, usually as the result of accidents, have been reported. One of the most spectacular was the accident at Poza Rica, Mexico, in November 1950. Large amounts of hydrogen sulfide escaped from a sulfur recovery plant over a period of about 20 min. Twenty-two persons died and 320 persons were hospitalized with symptoms of acute hydrogen sulfide poisoning. A number of animals also died. Severe respiratory tract irritation and loss of sense of smell was noted in the victims and the deaths which occurred were mostly of a central nervous system type with unconsciousness and dizziness. Pulmonary edema was noted in the victims (McCabe and Clayton, 1952).

E. SELF-POLLUTION

No discussion of the health effects of air pollution would be complete without devoting some comment to the self pollution of cigarette smoking. Results of carefully controlled scientific research accumulated over the past 10 to 15 yr clearly link cigarette smoking to lung cancer, emphysema, and chronic bronchitis. The attributable death rate in heavy smokers has been found to be four times that of light smokers. The death rate for coronary thrombosis was found to be higher than that for lung cancer when comparing non-smokers and heavy smokers. Hence, it would appear that heavy smoking causes a very large number of deaths from coronary artery disease, and the association with coronary heart disease is at least as important, from a public health point of view, as the association with lung cancer (Doll and Hill, 1964). A study of a religious group which practices abstinence from tobacco smoking demonstrated a preferential increased survival of the religious group over that of the simultaneous general population of the same area of the country. The authors concluded that an additional biologic cost of smoking is decreased life expectancy (Lemon and Kuzma, 1969).

Study of the various components of cigarette smoke which have been shown to have deleterious effects on human health has been quite extensive. Hence, it is known that tars and nicotines have specific detrimental effects. Smoking cigarettes equipped with efficient filters which lower the tar and nicotine content of smoke produces less pulmonary fibrosis and emphysema in male beagle dogs (Hammond et al., 1970). Auerbach et al. (1970), analyzing the same set of experiments, found more non-invasive bronchioalveolar tumors in the smoking group than in non-smoking dogs and significantly more in dogs smoking non-filter cigarettes than in dogs smoking filter tip cigarettes. In human males the thickness of myocardial arterial walls was found to be greater on the average in smokers than in non-smokers. The thickness also increased with age and with the number of cigarettes smoked per day (Auerbach et al., 1971). Allergy to tobacco protein or carbon monoxide in cigarette smoke may also contribute to the effect. Studies which have implicated cigarette smoking in a number of diseases have usually considered tar products and nicotine as the primary responsible agents. There has been very little attention paid, until recently, to carbon monoxide which is the largest continuous constituent of tobacco smoke.

Hammond (1966) compared coronary heart disease mortality ratios by age for current smokers of cigarettes by numbers of cigarettes smoked daily. Assuming that the mortality ratio for non-smokers is 1.0, the analysis, based on deaths from coronary heart disease over a 4 yr period among 441 000 men, showed that the rates were higher among men who smoke cigarettes than those who do not, and that the mortality ratios increase generally with increased intensity of cigarette smoking. Figure 2 illustrates that the highest mortality ratios are observed in the 45 to 54 yr old age group with the ratios decreasing with advancing age in each category. It should be noted that the 45 to 54 yr old age group is the same age group in the general population which has the highest mortality from coronary heart disease.

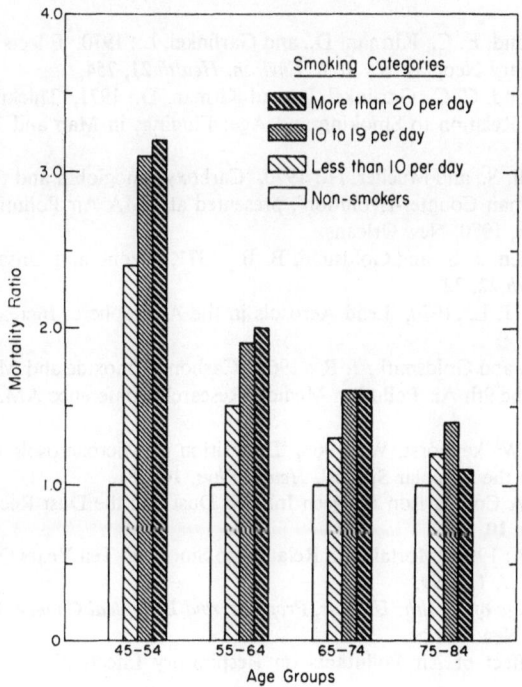

Fig. 2. Mortality ratios for coronary heart disease, by age groups, compared for smokers of cigarettes, by daily consumption. The mortality ratio for non-smokers is arbitrarily set at 1.0 for all ages. Data from Hammond (1966).

As with many other disease conditions we have discussed, simple cause and effect relationships do not answer the question of possible contributions of air pollution to the etiology of lung cancer. Direct and indirect relationships have not been fully explored and the co-factors that may be operating must be considered. The correlation between cigarette smoking and lung cancer is solid. However, the urban factor can easily be incriminated, and can, with appropriate statistical techniques, be separated out, leaving causation less than fully defined in the case of lung cancer and cigarette smoking. Air pollution is undoubtedly a major contributor to the urban factor. This leads naturally to the conclusions that air pollution is a major factor contributing to lung cancer.

There is no question that there are synergistic effects of cigarette smoking and air pollution. Neither air pollution nor cigarette smoking alone can be implicated as the sole element in the cardiopulmonary disease category. It is reasonable to assume that we may be seeing a single group of etiologic agents with variations in amounts and types of concentration and in modes of delivery of the irritant materials to the lungs.

References

Amdur, M. O.: 1969, 'Toxicologic Appraisal of Particulate Matter, Oxides of Sulphur, and Sulphuric Acid', 62nd Annual Meeting Air Pollution Control Association, *Proceedings Digest*, New York, 22.

Auerbach, O., Hammond, E. C., Kirman, D., and Garfinkel, L.: 1970, 'Effects of Cigarette Smoking on Dogs II. Pulmonary Neoplasms', *Arch. Environ. Health* 21, 754.

Auerbach, O., Hammond, C. C., Garfinkel, L., and Kirman, D.: 1971, 'Thickness of Walls of Myo-cardial Arteriole in Relation to Smoking and Age: Findings in Man and Dogs', *Arch. Environ. Health* 22, 20.

Ayres, S. M., Giannelli, S., and Mueller, H.: 1970, 'Carboxyhemoglobin and the Access to Oxygen: An Example of Human Counter-Evolution', presented at AMA Air Pollution Medical Research Conference, October, 1970, New Orleans.

Balchum, O. J., O'Brien, J. S., and Goldstein, B. B.: 1971, 'Ozone and Unsaturated Fatty Acids', *Arch. Environ. Health* 22, 32.

Chow, T. J. and Earl, J. L.: 1970, 'Lead Aerosols in the Atmosphere: Increasing Concentrations', *Science* 169, 577.

Cohen, S., Deane, M., and Goldsmith, J. R.: 1968, 'Carbon Monoxide and Myocardial Infarction', paper presented at the 9th Air Pollution Medical Research Conference AMA, Denver, Colorado, 1968.

Dautrebande, L. and Walkenhorst, W.: 1964, 'Deposition of Microaerosols in Human Lung with Special Reference to the Alveolar Spaces', *Health Phys.* 10, 981.

Davies, C. N.: 1964, 'A Comparison Between Inhaled Dust and the Dust Recovered From Human Lungs', *Health Phys.* 10, 1029.

Doll, R. and Hill, A. B.: 1964, 'Mortality in Relation to Smoking: Ten Years Observations of British Doctors', *Brit. Med. J.* 1, 1399.

Dubos, R.: 1959, *Mirage of Health: Utopias, Progress, and Biological Change*, Doubleday and Com-pany, Garden City, New York.

Ehrlich, R.: 1963, 'Effect of Air Pollutants on Respiratory Infection', *Arch. Environ. Health* 6, 638.

Fairbairn, A. S. and Reid, D. D.: 1958, 'Air Pollution and Other Local Factors in Respiratory Disease', *Brit. J. Prev. Social Med.* 12, 94.

Firket, J. (Secretary), 1931, 'Sur les causes des accidents survenus dans la vallee de la Meuse, los des brouillards de December, 1930', *Bull. Acad. Roy. Med. Belg.* 11, 683.

Frank, N. R.: 1964, 'Studies on the Effects of Acute Exposure to Sulphur Dioxide in Human Subjects', *Proc. Roy. Soc. Med.* 57 (Pt. 2), 1029.

Glasser, M. and Greenburg, L.: 1971, 'Air Pollution, Mortality, and Weather: New York City, 1960–1964', *Arch. Environ. Health* 22.

Goddard, R. F., Mercer, T. T., O'Neill, P. X. F., Flores, R. L., and Sanchez, R.: 1968, 'Output Characteristics and Clinical Efficacy of Ultrasonic Nebulizers', *J. Asthma Res.* 5, No. 4, 355.

Goldsmith, J. R. and Landaw, S. A.: 1968, 'Carbon Monoxide and Human Health', *Science* 162, 1352.

Goldstein, B., Buckley, R. D., Cardenas, R., and Balchum, O. J.: 1970, 'Ozone and Vitamin E', *Science* 169, 605.

Hammond, E. C.: 1966, 'Smoking in Relation to the Death Rates on One Million Men and Women', *Nat. Cancer Inst. Monograph* 19, 127.

Hammond, E. C., Auerbach, O., Kirman, D., and Garfinkel, L.: 1970, 'Effects of Cigarette Smoking on Dogs I. Design of Experiment, Mortality and Findings in Lung Parenchyma', *Arch. Environ. Health* **21**, 740.

Handler, Philip (Ed.): 1970, *Biology and the Future of Man*, Oxford University Press, New York.

Hatch, T. F. and Gross, P.: 1964, *Pulmonary Deposition and Retention of Inhaled Aerosols*, Academic Press, New York.

Hodgson, T. A., Jr.: 1970, 'Short-Term Effects of Air Pollution on Mortality in New York City', *Environ. Sci. Tech.* **4**, No. 7, 589.

Kerr, L. E.: 1956, 'Coal Workers' Pneumoconiosis', *Ind. Med. Surg.* **25**, 355.

Kerr, L. E.: 1969, 'The Occupational Pneumoconiosis of Coal Miners as a Public Health Problem', *Virg. Med. Monthly* **96**, 121.

Kerr, L. E.: 1970, 'Coal Workers' Pneumoconiosis in an Affluent Society', *Public Health Rep.* **85**, No. 10.

Langer, A. M., Selikoff, I. J., and Sastre, A.: 1971, 'Chrysotile Asbestos in the Lungs of Persons in New York City', *Arch. Environ. Health* **22**.

Lemon, F. R. and Kuzma, J. W.: 1969, 'A Biologic Cost of Smoking – Decreased Life Expectancy', *Arch. Environ. Health* **18**, 950.

Loomis, T. A.: 1968, *Essentials of Toxicology*, Lea and Febiger, Philadelphia.

Lowry, T. and Schuman, L. M.: 1956, '"Silo Fillers Disease", a Syndrome Caused by Nitrogen Dioxide', *J.A.M.A.* **162**, 153.

Mancuso, T. F.: 1970, 'Relation of Duration of Employment and Prior Respiratory Illness to Respiratory Cancer Among Beryllium Workers', *Environ. Res.* **3**, 251.

McCabe, L. C. and Clayton, G. D.: 1952, 'Air Pollution by Hydrogen Sulfide in Poza Rica, Mexico – An Evaluation of the Incident of November 24, 1950', *A.M.A. Arch. Ind. Hyg. Occup. Med.* **6**, 199.

Ministry of Health: 1954, 'Mortality and Morbidity During the London Fog of December, 1952', *Rep. Public Health Med. Subjects* **95**, H.M. Stationery Office, London.

Nadel, J. A., Salem, H., Tamplin, B., and Tokiwa, Y.: 1965, 'Mechanism of Bronchoconstriction During Inhalation of Sulphur Dioxide: Reflex Involving Vagus Nerves', *Arch. Environ. Health* **10**, 175.

National Academy of Sciences and National Academy of Engineering: 1969, 'Effects of Chronic Exposure to Low Levels of Carbon Monoxide on Human Health, Behavior, and Performance', NAS-NAE, Washington, D.C.

Pitts, J. N. and Metcalf, R. L. (Eds.): 1969, *Advances in Environmental Sciences*, Wiley-Interscience, New York.

Roueche, B.: 1947, 'The Fog' in *Eleven Blue Men*, Little, Brown and Company, New York.

Schrenk, H. H.: 1949, 'Air Pollution in Donora, Pennsylvania. Epidemiology of the Unusual Smog Episode of October, 1948', *Pub. Health Bull.* **306**, Federal Security Agency, Washington, D.C.

Schroeder, M. D., Hanover, N. H., and Brattleboro, V. T.: 1970, 'A Sensible Look at Air Pollution by Metals', *Arch. Environ. Health* **21**, 798.

Selikoff, I. J.: 1970, 'Asbestos Air Pollution', presented at AMA Air Pollution Medical Conference. October, 1970, New Orleans.

Sterling, T. D., Pollack, S. V., and Weinkam, J.: 1969, 'Measuring the Effect of Air Pollution on Urban Morbidity', *Arch. Environ. Health* **18**, 485.

Stokinger, H. E.: 1957, 'Evaluation of the Hazards of Ozone and Oxides of Nitrogen', *Arch. Ind. Health* **15**, 181.

Stokinger, H. E. and Coffin, D. L.: 1968, 'Biologic Effects of Air Pollutants', Ch. 13 in Stern, Arthur C. (Ed.) *Air Pollution*, 2nd Ed., Academic Press, New York.

Stokinger, H. E., Wagner, W. D., and Dobrogorski, O. J.: 1957, 'Ozone Toxicity Studies. III. Chronic Injury to Lungs of Animals Following Exposure at a Low Level', *Arch. Ind. Health* **16**, No. 6, 514.

U.S. Department of Health, Education, and Welfare: 1969, 'Air Quality Criteria for Sulfur Oxides: Summary and Conclusions', Consumer Protection and Environmental Health Service, Washington, D.C.

U.S. Department of Health, Education, and Welfare: 1970, 'Air Quality Criteria for Photochemical Oxidants', Environmental Health Service, National Air Pollution Control Administration, Washington, D.C., NAPCA Pub. AP-63.

Webster, W. S.: 1970, 'The Effect of Carbon Monoxide on Experimentally-Induced Atherosclerosis in the Squirrel Monkey', presented at AMA Air Pollution Medical Research Conference, October, 1970, New Orleans.

Winkelstein, W., Jr. and Kantor, S.: 1969, 'Stomach Cancer: Positive Association with Suspended Particulate Air Pollution', *Arch. Environ. Health* **18**, 544.

Winkelstein, W. and Gay, M. E.: 1971, 'Suspended Particulate Air Pollution Relationship to Mortality from Cirrhosis of the Liver', *Arch. Environ. Health* **22**, 174.

Yuen, T. T. H. and Sherwin, R. P.: 1971, 'Hyperplasia of Type 2 Pneumocytes and Nitrogen Dioxide (10 ppm) Exposure: A Quantitation Based on Electron Photomicrographs', *Arch. Environ. Health* **22**, 178.

Bibliography

Carnow, B. W., Senior, R. M., and Karsh, R. (Discussants), Wessler, S. and Avioli, L. V. (Editors): 1970, 'The Role of Air Pollution in Chronic Obstructive Pulmonary Disease', *J.A.M.A.* **214**, No. 5, 894.

Cohen, S. I., Perkins, N. M., Ury, H. K., and Goldsmith, J. R.: 1971, 'Carbon Monoxide Uptake in Cigarette Smoking', *Arch. Environ. Health* **22**, 55.

Davies, C. N. (Ed.): 1961, *Inhaled Particles and Vapours*, Proceedings of an International Symposium Organized by the British Occupational Hygiene Society, Oxford, March 1960, Symposium Publications Division, Pergamon Press, New York.

Engel, S.: 1964, 'Comparative Anatomy and Pulmonary Air-Cleansing Mechanisms in Man and Certain Experimental Animals', *Health Phys.* **10**, 967.

Ferris, B. G.: 1969, 'Chronic Low-Level Air Pollution: Use of General Mortality, and Chronic Disease Morbidity and Mortality to Estimate Effects', *Environ. Res.* **2**, 79.

Gafafer, W. M. (Ed.): 1964, *Occupational Diseases, A Guide to Their Recognition*, Public Health Service Publication No. 1097, Washington, U.S. Govt. Printing Office.

Goldsmith, J. R.: 1968, 'Effects of Air Pollution on Human Health', Ch. 14 in Stern, Arthur C. (Ed.) *Air Pollution*, 2nd., Academic Press, New York.

Goldsmith, J. R.: 1969, 'Air Pollution Epidemiology', *Arch. Environ. Health* **18**, 516.

Gross, P.: 1964, 'The Processes Involved in the Biologic Aspects of Pulmonary Deposition, Clearance, and Retention of Insoluble Aerosols', *Health Phys.* **10**, 995.

U.S. Department of Health, Education, and Welfare: 1969, 'Air Quality Criteria for Particulate Matter: Summary and Conclusions', Consumer Protection and Environmental Health Service, Washington, D.C.

U.S. Department of Health, Education and Welfare: 1969, 'Guidelines for the Development of Air Quality Standards and Implementation Plans', Consumer Protection and Environmental Health Service, National Air Pollution Control Administration, Washington, D.C.

U.S. Department of Health, Education, and Welfare: 1970, 'Air Quality Criteria for Photochemical Oxidants', Environmental Health Service, National Air Pollution Control Administration, Washington, D.C., NAPCA, Pub. AP-64.

Winkelstein, W., Jr. and Kantor, S.: 1969, 'Respiratory Symptoms and Air Pollution in an Urban Population of Northeastern United States', *Arch. Environ. Health* **18**, 760.

Wolozin, H. (Ed.): 1966, *Tne Economics of Air Pollution, A Symposium*, W. W. Norton & Co., New York.

EFFECTS OF AIR POLLUTANTS ON VEGETATION

SAMUEL N. LINZON

Chief, Phytotoxicology Section, Air Management Branch, Ontario Department of the Environment, Ont., Canada

1. Introduction

Plant response to air pollutants has demonstrated in a striking manner the adverse effects of an industrialized society. The number and kinds of air pollutants added to the environment have increased dramatically in recent decades. Damage to vegetation from air pollution can be considerable and it is estimated that injury to agricultural and forestry crops in the United States amounts to hundreds of millions of dollars annually (Brandt and Heck, 1968). This estimate does not take into account economic losses due to growth suppression, delayed maturity and future yield values of some plants. The prompt detection of plant damage resulting from air pollution is extremely important since plants will often display symptoms of injury before effects are discernible in man.

In this chapter the effects of various air pollutants on vegetation are outlined. Almost one hundred references are given covering the period 1893 to 1971. The major pollutants are treated first, followed by the lesser known contaminants with respect to vegetation. Information on symptoms of injury, dosages required to cause injury, sensitivity and tolerance of various plant species, and predisposing environmental and other factors are discussed. Consideration is also given to methods of investigating and preventing the effects of air pollutants on vegetation.

2. Effects of Various Pollutants

A. SULPHUR DIOXIDE

There is reference to the deleterious effects of SO_2 on vegetation dating back to more than 100 yr ago in Europe. In the United States the Experiment Station of the Agricultural College of Utah published a bulletin in 1903 which described the effects of smelter smoke on Utah agriculture (Widtsoe, 1903). Litigation over vegetation damage caused by smelter funes took place in the early 1900's at Ducktown, Tennessee, and Anaconda, Montana (Swain, 1949). About a decade later injunctions against smelters and investigations took place at Salt Lake City, Utah, and Selby, California. In the third decade of this century, an international problem arose when smelter fumes emitted by the Consolidated Mining and Smelting Company at Trail, British Columbia, travelled down the Columbia River Valley to damage forests in Stevens County in the State of Washington. Comprehensive investigations were carried out for about 10 yr, which resulted in the publication of a book by the National Research Council of Canada (Katz *et al.*, 1939). Investigations in the Sudbury district of

McCormac (ed.), Introduction to the Scientific Study of Atmospheric Pollution, 131–151. All Rights Reserved.
Copyright © 1971 by D. Reidel Publishing Company, Dordrecht-Holland.

Ontario started in the 1940's (Katz, 1949) and are still continuing (Linzon, 1958; Linzon, 1971).

A1. *Symptomatology of Sulphur Dioxide Injury on Vegetation*

Sulphur dioxide enters leaves mainly through the stomata. The toxicity of this gas to the mesophyll cells of the leaves is primarily due to its reducing properties. Injury to the leaves has been classified into two types: acute or chronic. Acute injury is caused by absorption of high concentrations of SO_2 in a relatively short time. This results in a rapid accumulation of sulphite which is toxic to the metabolic processes taking place in the mesophyll cells. Chronic injury is caused by a long term absorption of SO_2 at sub-lethal concentrations. The sulphite formed is oxidized to sulphate at about the same rate that the gas is absorbed. When sulphate accumulates beyond a threshold value that the plant cells can tolerate, chronic injury occurs. It is estimated that sulphate is about 30 times less toxic than sulphite (Thomas, 1961).

Acute injury on broad leaves takes the form of bifacial lesions, which usually occur between veins, and is often more prominent towards the petiole. The injury is local. The metabolic processes are completely disrupted in the necrotic or dead areas, with the surrounding leaf tissue remaining green and functional. The tissue on either side of the veins is extremely resistant. In some cases, injury can occur on the margins of the leaves. Young leaves rarely display necrotic markings, whereas fully expanded leaves are most sensitive to this type of SO_2 injury. The oldest leaves are moderately sensitive. In monocotyledonous leaves the injury can occur at the tips and in lengthwise areas between the main veins. In conifers, acute injury usually appears as a bright orange-red tip necrosis on current-year needles, often with a sharp line of demarcation between the injured tips and the normally green bases. Occasionally the injury may occur as bands in apical, medial, or basal locations on the needles.

Chronic injury becomes manifest as a yellowing or chlorosis of the leaf, sometimes from lower to upper surfaces on broad leaves (Brisley *et al.*, 1959). Occasionally only a bronzing or silvering will occur on the under surface of the leaves (Barrett and Benedict, 1970). The rate of metabolism is reduced in leaves displaying chronic injury. In conifers, chronic injury on older needles appears as a yellowish-green color, then changes to reddish-brown, starting at the tips and developing basipetally (Linzon, 1969).

The anatomical effects of acute injury may be observed under the microscope (Katz and Shore, 1955). In the initial stages of injury there are water-soaked, flaccid areas of diffuse greyish-green coloration. The chlorophyll appears to have diffused from the chloroplasts into the cytoplasm. This is followed by desiccation and shrinkage of the affected cells. The green pigments are decomposed, and the affected leaf area assumes a bleached, ivory, tan, orange-red, reddish-brown, or brown appearance, depending upon the plant species, the time of the year, and the weather conditions.

Early investigators studying the effects of SO_2 on vegetation concluded that 'invisible injury', or physiological disturbance, or effects on growth on yield of plants, could occur in the absence of chlorotic or necrotic markings (Wislicenus, 1901;

Stoklasa, 1923). This theory has been refuted by several investigations carried out in North America in the 1930's (Hill and Thomas, 1933; Swain and Johnson, 1936; Setterstrom *et al.*, 1938; Katz *et al.*, 1939). However, recent studies in Europe have revived the controversy, by reporting that reductions in growth and yield occurred without visible markings on certain plants exposed to atmospheric SO_2 in the field (Bleasdale, 1959; Guderian and Strattman, 1962). It is quite possible that other pollutants in the ambient air acted in concert with the low concentrations of SO_2 to cause the observed effects.

A2. *Dosages of Sulphur Dioxide Required to Cause Injury*

The most publicized criterion is that injury to vegetation will rarely occur if the 8 hr average concentration of SO_2 does not exceed 0.25 ppm SO_2. Field studies and fumigation experiments support this criterion (Thomas and Hill, 1937; Katz *et al.*, 1939; Linzon, 1958; Dreisinger, 1965). Other potentially damaging concentrations to vegetation are 0.35 ppm for 4 hr; 0.55 ppm for 2 hr and 0.95 ppm for 1 hr. An average annual SO_2 concentration of 0.03 ppm or over in the vicinity of a few stationary sources usually indicates that chronic or acute vegetation injury occurred during the growing season from higher potentially damaging concentrations.

A3. *Sensitivity and Tolerance of Various Plant Species and Varieties to Sulphur Dioxide*

Different plant species and varieties, and even individuals of the same species, may vary considerably in their sensitivity or tolerance to SO_2. Susceptibility lists have been made by several investigators. These lists should be used only as a guide since variations can occur because of differences in geographical location, climate, and plant stage of growth and maturation.

O'Gara (1922) listed the sensitivity of about 100 plants to SO_2 as determined in fumigation experiments which were validated to some extent by Thomas and Hendricks (1956). Alfalfa, barley, endive, and cotton were most sensitive with privet being 15 times more tolerant to SO_2.

Zahn (1961) grouped crops into three resistance groups according to their tolerance limits. Clover-type fodder plants were most sensitive to SO_2; wheat, leafy vegetables (excluding cabbage), beans, strawberries and roses were moderately sensitive; and roots and cabbage were least sensitive.

Katz *et al.* (1939) published lists of relative susceptibility of plants to SO_2 in British Columbia based on field observations and fumigation experiments. Larch, birch, ninebark, alfalfa and lettuce were most sensitive, whereas red cedar, silver maple, spirea, field corn, and asparagus were most tolerant.

At Winnipeg, Manitoba, Linzon (1965) found the following order of susceptibility to SO_2. Bur oak, Manitoba maple, and trembling aspen were most susceptible, white elm and choke cherry intermediate, and balsam poplar and green ash most resistant.

At Sudbury, Ontario, Dreisinger (1965) reported on the susceptibility of cultivated plants and native forest trees to SO_2 based on observations in the field. Buckwheat, red clover, trembling aspen, jack pine, eastern white pine, white birch and bracken fern were most sensitive, whereas cabbage, corn, cedar, spruce and maple were most tolerant.

Zimmerman and Hitchcock (1956) determined the comparative susceptibility of a number of plants to SO_2 in fumigation experiments. Chicory, alfalfa, geranium, buttonbush and eggplant were most sensitive, whereas, Jerusalem cherry, tulip, milo maize, ixora, and corn (field and sweet) were most tolerant.

Weeds are as sensitive to air pollutants as are commercial plants. Knowledge of their sensitivity to SO_2 is useful in field studies. Benedict and Breen (1955) selected 10 weeds which occur commonly throughout the United States and determined their sensitivity to SO_2 in fumigation experiments. Chickweed was most sensitive; mustard, annual bluegrass, sunflower, Kentucky bluegrass, pigweed and cheeseweed were intermediate in sensitivity; and lambs-quarter, dandelion and nettle-leaf goosefoot were tolerant.

A4. *Economic Effects of Sulphur Dioxide on Forest Growth*

Forests in the Sudbury district of Ontario have been subjected to S fumes emanating from the roasting and smelting of Ni-Cu ores since 1888. By 1950, three large smelters were discharging annually over 2 million tons of SO_2 into the surrounding atmosphere.

A study was conducted over a number of years on the chronic effects of SO_2 on yield, growth, and survival of white pine in the Sudbury area. Based on the degree of injuries exhibited by the trees, and on air sampling records of SO_2, the Sudbury area was segregated into three fume zones: Inner, Intermediate and Outer (Linzon, 1966). In the Inner Fume Zone, an area of about 720 mile^2, white pine trees displayed severe foliar injuries which resulted in reduced radial and volume growth and in excessive mortality. Over three times as many trees died annually in the Inner Fume Zone than in a control area remote from the sources of SO_2 fumes. An estimate was made of the loss in income to the producers of wood in the Sudbury area. For white pine alone, which represents only 7.6% of the total productive forest area in parts of the Sudbury district, a loss of 117 000 dollars was estimated to occur annually in the Inner Fume Zone (Linzon, 1971).

A5. *Environmental and Other Factors*

Environmental factors which are conducive to optimum plant growth are usually the same factors which abet SO_2 injury. These factors include sunlight, moderate temperature, high relative humidity, wind, and adequate soil moisture. In addition, time of day and season, and plant factors such as genotype, nutrition, stage of growth, and tissue maturation determine the sensitivity of a particular species to SO_2 injury.

Vegetation is most susceptible to SO_2 during the active growth months of June, July, and August. If the environmental factors and the growth stages of the plants are not conducive to injury, the plants will escape injury even in the presence of potentially damaging concentrations of SO_2.

Most investigators have shown a direct relationship between open stomata and the absorption of SO_2 and subsequent leaf injury. When stomata are closed, either at night because of darkness or during the day because of other factors, plants are more resistant to SO_2. It has been reported that the potato plant is as equally sensitive at night or during the day because their stomata do not close at night.

The main factors responsible for stomatal opening are sunlight, cell turgidity, and temperature. A sequence of events could be illustrated as follows: Light→Increased photosynthesis→Decreased CO_2 content

phosphorylase (high pH enzyme)

→Increased cell pH→starch⇌sugar→Increased cellosmotic pressure

low pH enzyme

→Increased turgidity of guard cells due to water entering→ Open stomata.

The closing of the stomata would arise from the reverse of the above events. Zimmerman and Hitchcock (1956) found no relation between the number of stomata per unit leaf area of different parts and relative susceptibility to SO_2.

Recent work indicates that biochemical factors apart from those controlling stomatal action also are related to the sensitivity of plants to air pollutants.

B. OXIDANTS

Ozone and PAN (peroxyacetyl nitrate) are the main phytotoxicants in the Los Angeles type of oxidant smog now plaguing many urban areas. Automobile exhaust is the major contributor of the primary substances (NO_x and hydrocarbons) in the photochemical reaction producing the secondary toxic pollutants.

Injury to vegetation from photochemical air pollution was first observed in the Los Angeles area in 1944 (Middleton *et al.*, 1950). In 1958 (Heggestad and Middleton, 1959) atmospheric O_3 was found to be the cause of 'weather fleck' of tobacco which had caused extensive injury in northeastern U.S.A. and in Ontario, Canada, starting in the early 1950's. Artificial O_3 fumigation experiments conducted in the laboratory to duplicate symptoms of injury observed on eastern white pine trees were started in Canada in 1959 (Linzon, 1966) and in the United States in 1961 (Berry and Ripperton, 1963).

Ozone is estimated to be responsible for 85 to 90% of the total oxidizing potential of photochemical air pollution (Taylor, 1968). Other known phytotoxicants in the smog complex are NO_2, SO_2, and the PAN's. Ozone is probably responsible for injuring more plant species than any other phytotoxic air pollutant. Forty plant species have been listed as being relatively sensitive to O_3 (Hill *et al.*, 1970).

C. OZONE

C1. *Symptomatology of Ozone Injury*

Ozone injury can result in four general types of lesions. These include pigmentation, chlorosis, bleaching, and necrosis (Hill *et al.*, 1961, 1970).

Pigmented lesions are usually due to thickening and pigmentation of cell walls (Ledbetter *et al.*, 1959). Palisade mesophyll cells are most susceptible to O_3 injury with the brown, black, red or purple lesions primarily occurring on the upper surfaces of the leaves.

Chlorotic lesions usually occur in the upper leaf surfaces because of the injury to palisade cells. The lesions may coalesce to present a mottled appearance. Monocotyledonous blade-like leaves which do not possess differentiated mesophyll tissue may develop the chlorotic mottling on either leaf surface. Pine needles usually have the chlorotic flecks start on stomatal faces.

Bleaching, which usually occurs on upper leaf surfaces is due to the collapse of the palisade cells and sometimes the upper epidermal cells.

Necrosis occurs when the leaf tissue is killed and usually the necrosis is bifacial extending through both leaf surfaces. On monocotyledonous and coniferous leaves the necrosis usually takes the form of a tip injury. The colors of the necroses range from white through tan to orange-red depending on the plant species. Ozone injury is usually interveinal, although lesions frequently concentrate along the sides of main veins.

Ozone enters the leaf through the stomata and preferentially injures the palisade cells (Taylor, 1970). Ozone injury to broad leaves displays a definite pattern relating to the development of functional stomata (Menser et al., 1963). The youngest leaves are resistant and with expansion become susceptible at their tips. With increasing maturity the leaves become successively susceptible at middle and basal portions, and with complete maturation the leaves become resistant again.

Needle injuries attributed to atmospheric oxidants occurring in white pine forests in the east and ponderosa pine forests in the west have been the subject of intensive investigations in recent years. These injuries have been described and have been given names as O_3 injury of eastern white pine (Linzon, 1967; Costonis and Sinclair, 1969), chlorotic dwarf of eastern white pine (Dochinger, 1968), and chlorotic decline of ponderosa pine (Miller et al., 1963). A physiogenic disease affecting eastern white pine is described as semimature-tissue needle blight (SNB) with the injury development resembling part of the symptom syndrome caused by O_3 (Linzon, 1966). The symptoms of these different pine needle injuries as seen in the forest range from minute silvery flecks through chlorotic mottle to tip necrosis.

The subcellular effects of O_3 involve two stages (Thomson et al., 1966). First, a granulation occurs in the chloroplast stroma and, secondly, there is a general disruption of the cells with the cellular contents aggregating in the center of the cell. Linzon (1967) found the microscopical symptoms of O_3 injury and SNB of eastern white pine to be dissimilar.

In addition to visible injury, growth suppression may result from the effects of oxidants in decreasing photosynthesis and changing cell membrane permability (Todd, 1958; Dugger et al., 1966; Hill and Littlefield, 1969).

C2. *Dosages Required to Cause Injury*

Ambient air concentrations of 3 pphm oxidant and over for several hours in forests in Canada have been found by this writer to cause O_3 injury symptoms on current-year needles of sensitive eastern white pine trees. Oxidant concentrations of 15 pphm for approximately 2 hr per day for about 2 months in Southern California have caused chlorotic decline symptoms on 1 yr old needles of susceptible ponderosa pine trees (Miller *et al.*, 1969).

Hill *et al.* (1970) report that typical lesions may be produced on tobacco leaves by concentrations of O_3 as low as 5 pphm for 4 hr, and sensitive varieties of alfalfa, spinach, clover, oats, radish, sweet corn, and bean have been injured by 10 to 12 pphm for 2 hr.

Heck and Tingey (1970) have projected the minimum O_3 concentrations which will produce, for short term exposure, 5% injury to sensitive vegetation grown under optimum conditions. These time concentrations are: 15 pphm for 0.5 hr, 10 pphm for 1 hr, 7 pphm for 2 hr, 5 pphm for 4 hr and 3 pphm for 8 hr.

C3. *Sensitivity and Tolerance of Plant Species*

There is considerable variation in sensitivity to O_3 injury amongst plant species and varieties. In addition to interspecific variation there is inherent intraspecific variation among the individuals of the same plant species. Further, on eastern white pine, induced symptoms of injury not only vary from tree-to-tree, but also from needle-to-needle, fascicle-to-fascicle, and branch-to-branch on the same tree.

For experimental work, sensitive plants such as Bel-W3 tobacco, pinto bean, and Norland potato are extremely useful as indicators of the presence of low concentrations of oxidants. Hill *et al.* (1970) listed 20 crop plants and 20 trees, shrubs and ornamentals which are relatively sensitive to O_3. Some of these plants were alfalfa, barley, bean, muskmelon, oat, onion, potato, spinach, tobacco, tomato, wheat, grape, petunia, eastern white pine, and ponderosa pine.

C4. *Environmental and Other Factors*

Various environmental factors predispose plants to O_3 injury (Heck, 1968). These include poor nutrition, low light, high relative humidity, and moderately high temperatures. Stomatal closure due to moisture stress can protect plants from O_3 injury (Macdowall, 1965). Low carbohydrate content within the plant increases sensitivity, and Lee (1965) could protect plants from O_3 injury by increasing sugar levels. Leone *et al.* (1966) found greatest sensitivity of tobacco plants to O_3 injury when they were supplied with levels of N conducive to optimum growth.

D. OZONE AND SULPHUR DIOXIDE SYNERGISM IN CAUSING INJURY TO VEGETATION

Menser and Heggestad (1966) reported injuring susceptible tobacco plants with a mixture of 2.7 pphm O_3 and 24.0 pphm SO_2 for 2 hr in which the two gases acted synergistically. The plants were not injured when exposed to either gas alone at the concentrations used in the combined-gas experiment. Applegate and Durrant (1969) reported the synergistic action of O_3 and SO_2 on peanuts. Dochinger *et al.* (1970) utilized a mixture of 10 pphm O_3 and 10 pphm SO_2 on eastern white pine trees to produce symptoms similar to those caused by the chlorotic dwarf disease. It was suggested that the two pollutants acted synergistically since the degree of injury produced by either of the two pollutants alone at the same concentrations was much less than that produced by the mixture. Houston and Stairs (1970) injured sensitive eastern white pine clones with a mixture of 2.5 pphm SO_2 and 5 pphm O_3 in a 6 hr fumigation. The plants were injured when treated with 2.5 pphm SO_2 alone, but were not injured when treated with 5 pphm O_3 alone.

Additional research is required to explore and explain the action of mixtures of air pollutants on vegetation. Ozone is a natural constituent of our air resource. The concentrations fluctuate diurnally in urban, rural, and forested areas, and can persist at high levels over large areas under certain meteorological conditions. More studies are needed to determine the effects on vegetation of mixtures of O_3 and other pollutants as SO_2 and fluoride.

E. PEROXYACETYL NITRATE (PAN)

E1. *Symptomatology of PAN injury*

The typical symptoms of PAN injury are glazing and bronzing of lower leaf surfaces. Leaves of sensitive plant species first develop a slightly oily or waxy appearance about 2 to 3 hr after exposure and the glazed symptom develops gradually (Taylor and MacLean, 1970). The glazing is due to the collapse of mesophyll cells with large air pockets taking their place (Middleton *et al.*, 1950). Membrane permeability is affected and cellular contents leak into the intercellular spaces. Bronzing develops after 2 to 3 days. Glazing and bronzing have been observed also on the upper surfaces of tomato, petunia, and tobacco leaves.

If concentrations of PAN are high severe injury may occur and take the form of distinct transverse bands across the leaf (Taylor, 1970). A single band is produced by a single exposure but several bands may develop from successive exposures. Leaf tissue of a specific physiological stage is sensitive to injury with a younger or older tissue on the same leaf not affected during a fumigation. This type of injury with bands occurring at different levels in leaves of different maturity is most noticeable on monocotyledonous blade-like leaves and on compound leaves of plants as the tomato and potato.

PAN suppresses photosynthesis (Dugger *et al.*, 1965) which may account for the growth reductions observed in tomato and bean plants when exposed to low concentrations of PAN for several days.

E2. *Dosages Associated with PAN Injury*

Sensitive plants as tomato and lettuce have been severely injured by 15 to 20 ppb of PAN in a 4 hr exposure (Taylor and MacLean, 1970). Visible symptoms of PAN injury were observed on sensitive petunia plants when ambient air concentrations did not exceed 14 ppb. The highest concentration of ambient air PAN recorded at Riverside, California, was 58 ppb (Taylor, 1970).

E3. *Sensitivity and Tolerance to PAN Injury*

Taylor and MacLean (1970) listed the sensitivities of selected plant species to PAN. Pinto bean, Swiss chard, chickweed, dahlia, annual bluegrass, lettuce, mustard, little-leaf nettle, oat, petunia, and tomato were most sensitive. Those listed as intermediate in sensitivity included alfalfa, barley, beet, carrot, cheeseweed, sour dock, lambs-quarters, soybean, spinach, tobacco, and wheat. Tolerant plants included azalea, lima bean, begonia, broccoli, chrysanthemum, corn, cotton, cucumber, onion, periwinkle, radish, sorghum, and touch-me-not. The tolerant plants could resist injury in 2 hr exposures to 75 to 100 ppb of PAN.

E4. *Environmental and Other Factors*

Plants are more sensitive to PAN injury during periods of high light intensity. A minimum of light is required before, during, and after exposure to PAN for visible symptoms to develop. Approximately 3 hr of sunlight is required before and after a PAN fumigation for sensitive plants to respond to PAN. The reaction is reversible if an exposure period to PAN is followed immediately by a dark treatment of up to 3 hr (Taylor, 1970).

F. NITROGEN OXIDES

There are several kinds of oxides of N, but only NO and NO_2 occur in the atmosphere in significant concentrations. Nitric oxide is formed from high temperature combustion processes, and is oxidized in the air to NO_2. Atmospheric concentrations of NO_x and NO_2 rarely exceed 1.0 ppm and 0.5 ppm, respectively (State of Calif., 1966). The maximum concentrations recorded in Los Angeles by 1964 were 3.7 ppm for NO_x and 1.3 ppm for NO_2.

Nitrogen dioxide is the only oxide of N found to injure vegetation at the concentrations which occur in ambient air. Taylor (1970) reported that concentrations of NO of over 25 ppm failed to injure several plant species.

Nitrogen dioxide produces plant injury symptoms similar to those caused by SO_2. Typically, white or light tan colored lesions are produced between the main veins of the leaf. Less severe injury is manifested as a grey discoloration of the lower leaf surface (Taylor, 1970).

Nitrogen dioxide can injure the same plants as O_3, and in the same physiological tissue, but the injury symptoms are different, and approximately ten times as much NO_2 is required (Taylor and Eaton, 1966).

Environmental conditions significantly affect the dosages of NO_2 required to cause injury on plant species (Taylor and MacLean, 1970). Under light intensities equivalent to full sunlight about 6.0 ppm of NO_2 for 2 hr is required to injure sensitive species as pinto bean, tomato and cucumber. Sensitivity to NO_2 increases in low light because nitrate reduction processes in the plant are not active. Thus, under extremely low light intensity the same sensitive plant species may be injured when exposed to

2.5–3.0 ppm of NO_2 for 2 hr. Low concentrations of NO_2 (less than 0.5 ppm) inhibited the growth of pinto bean and tomato plants under continuous exposure for 10 to 22 days (Taylor and Eaton, 1966). No visible injury developed at these low concentrations, and indeed, the green color of the leaves was found to be intensified due to increased chlorophyll content.

Taylor and MacLean (1970) listed several plant species according to their sensitivity and tolerance to NO_2. Azalea, pinto bean, brittlewood, hibiscus, head lettuce, mustard, sunflower, and tobacco were considered sensitive. Some of the more tolerant plant species were asparagus, bush bean, lambs-quarters and pigweed.

G. FLUORIDES

Fluoride injury to vegetation was observed in Germany late in the 19th century. Mayrhofer (1893) reported injury in the vicinity of superphosphate fertilizer industries. In the United States the effects of fluoride on vegetation became a serious problem in the 1940's near aluminum industries (de Ong, 1946) and in the mining and processing of phosphate deposits in Florida and Tennessee (MacIntyre, 1949). Studies in Spokane County, Washington State, in the early 1950's showed ponderosa pine to be severely damaged by fluoride in an area 3.5 miles wide by 12 miles long (Adams et al., 1952). In Ontario, Drowley et al. (1963) reported soft suture injury to peaches in the early 1960's caused by fluorides emitted from a tile plant. Also in Ontario, Linzon (1970) reported results of fluoride-vegetation studies carried out in 1969 in the vicinity of fertilizer, fiberglas, and aluminum industries.

The term fluorides applies to a number of fluorine compounds which may occur in the atmosphere. These include gases as HF, SiF_4, and H_2SiF_6, and solids such as Na_3AlF_6, AlF_3, fluorapatite, CaF_2 and NaF. Instruments have not been developed which can differentiate these individual compounds which occur in minute concentrations in the atmosphere.

G1. *Symptomatology of Fluoride Injury on Vegetation*

Fluorides enter leaves primarily through the stomata, and are translocated towards the margins of broad leaves and to the tips of monocotyledonous and coniferous needles via the transpirational stream (Jacobson et al., 1966). Gaseous fluorides are more readily absorbed than particulate fluorides. When the accumulated fluoride concentrations exceed certain threshold levels, which varies with different plant species, leaf injury occurs. The injury may take the form of chlorosis, before necrosis occurs. Injured plants usually display necrotic tips or margins on leaves, with a darker zonate line delineating the margin of the injured area.

Fluoride injury starts as a gray or light-green, water-soaked discoloration of leaf tissue which changes to a tan or reddish-brown color (Treshow and Pack, 1970).

Fluoride injury on plants may be chronic or acute. In chronic injury, fluorides are absorbed slowly and move distally in the leaf. The action is systemic, and the lesions formed continue to enlarge as long as the plants are being subjected to atmospheric fluorides. In acute injury, inactivation and translocation cannot proceed as rapidly as the rate of absorption of high concentrations of atmospheric fluorides, which results in scattered interveinal lesions.

On gladiolus leaves the necrotic tips are light brown in color with a dark brown margin. The necrotic area is separated from the uninjured portion of the leaf by a narrow band of chlorotic tissue (Hitchcock et al., 1962). On corn, sorghum, and other grasses, the symptoms first appear as scattered chlorotic flecks at the tips and upper margins of middle-aged leaves. With intensification of injury the tips and margins become necrotic (Hitchcock et al., 1964).

On broad leaves the early symptoms take the form of a chlorosis of the leaf tip, which later extends down the margin and inward between the veins leaving a green 'christmas-tree' design on a chlorotic background (Weinstein and McCune, 1970).

Growth of the leaf during exposure can result in a crinkled appearance. The marginal chlorosis may become necrotic. Necrotic tips may fall out leaving a cordate-shaped leaf. The breaking away of necrotic margins produces a typically tattered edge on fluoride-affected leaves.

Coniferous needles display a light brown to reddish-brown tip necrosis which progresses basipetally. The proximal portion of the injured area is darker in color and can be used to distinguish repeated fumigations (Weinstein and McCune, 1970). Needles are most sensitive when they are elongating in the spring, with fully expanded needles and needles formed in previous years being increasingly more resistant to fluoride injury (Treshow and Pack, 1970).

Flower petals are rarely injured by atmospheric fluorides. Gladiolus plants displaying severe necrosis of leaves and bracts will produce unmarked spikes of flowers. Certain fruits may be injured. Soft suture of peach is characterized by a premature reddening and softening of the flesh along the suture line toward the stylar end of the fruit. It occurs before pit hardening when calcium demands are greatest (Treshow and Pack, 1970). In Ontario, soft suture of peaches was found to occur when atmospheric levels of fluoride exceeded 44 μg collected on 100 cm^2 of filter paper over a 30 day period (Drowley et al., 1963). In another study in Ontario in the vicinity of a fiberglas manufacturer, where much higher levels of fluoride were encountered in the air, silver maple trees displayed severe necrosis on the leaves and on the wings and seed casings of fruit (Linzon, 1970).

Fluorides inhibit photosynthesis, with the impairment initially being greater than the degree of visible leaf injury. With continued fumigation, the decrease in the rate of photosynthesis paralleled the increase in leaf tissue necrosis (Thomas, 1958). Fluorides inhibit enzymes in vitro, with a well-known example being enolase, an enzyme required in the glycolytic pathway of plant respiration. However, in vivo, the effects of fluoride on respiration are not clear, with both stimulation and inhibition reported (Weinstein and McCune, 1970).

Solberg et al. (1955) examined the subcellular histological effects of fluorides on pine needles. He observed that phloem and xylem parenchyma cells enlarged greatly and that mesophyll cells exhibited granulation, vacuolation, and finally total collapse. The resin duct epithelial cells were observed to enlarge greatly, to the extent that duct canals were frequently occluded.

G2. *Dosages Required to Cause Injury*

There is considerable variability in response to atmospheric fluorides by various plant species and varieties. McCune (1969) reviewed the literature for threshold concentrations and found that sensitive plant species were affected by about 0.4 to 1 μg F m^{-3} (0.82 μg HF m^{-3} is equivalent to 1 ppb) for time periods of from one to several weeks and that more resistant species required about 10 times as much fluoride to be injured. The above dosages were determined from continuous fumigation experiments. However, in the field, plant life would be subjected to intermittent fumigations. The effects evoked by plants would be different if the plants were exposed to the same total concentrations, but in one case for a continuous period of time, and in the other for intermittent periods of time.

G3. *Sensitivity and Tolerance of Plant Species*

Weinstein and McCune (1970) listed the relative sensitivity of 75 cultivated and native plants to injury by HF. This list which was derived from the works of various investigators was prepared only as a guide because inconsistencies may occur due to varietal, environmental and geographical factors. Gladiolus, Chinese apricot, Italian prune, plum, tulip, iris, St. Johnswort, sweet corn, grape, and developing needles of ponderosa pine and eastern white pine were placed in the sensitive category. Also placed in the sensitive category were smartweed and lamb's quarters, which in Ontario near fluoride sources are rarely injured. This illustrates the effects of environmental and geographical factors. Plants placed in the resistant category were petunia, apple, alfalfa, cotton, tobacco, bean, celery, cucumber, squash, cabbage, cauliflower, eggplant, and privet.

   The placing of plant species in the different categories of sensitivity to fluorides is based on several expressions of response. These include the threshold concentrations required to induce injury, the degree of injury produced by various dosages, and the concentrations of fluoride accumulated within the leaves. Chemical analysis of leaf samples can be a useful tool in this respect. Certain sensitive gladiolus varieties can display leaf injury with only 20 ppm fluoride in their leaves (Jacobson *et al.*, 1966), whereas bitternut hickory can concentrate up to 1000 ppm fluoride in its leaves without showing any visible injury (Linzon, 1970).

   Farm animals may develop the disease chronic fluorine toxicosis by feeding on fluoride-contaminated forage crops over an extended period of time. Shupe *et al.* (1970) listed cattle as most sensitive, followed by sheep, horses, swine, and poultry in decreasing sensitivity. Suttie (1969) recommended that animal feed should not be in excess of 40 ppm fluoride (dry weight basis) on an annual average based on monthly sampling, in order to avoid the adverse effects of the disease.

G4. *Environmental and Other Factors*

Zimmerman and Hitchcock (1956) found that plants grow under optimum moisture conditions were most sensitive to fluorides. McCune *et al.* (1966) found plants to be more sensitive when K, P, or Mg were deficient. Pack (1966) and MacLean *et al.* (1969) found increased injury on tomato plants grown under conditions of deficient Ca. Brennan *et al.* (1950) found that deficient and superoptimal levels of N reduced fluoride toxicity in tomato, whereas optimum levels of N enhanced toxicity.

H. CHLORINE

Chlorine is not widespread in the atmosphere, and is usually found confined to the immediate area sounding its source. Sources of Cl emissions include accidental spills from tank cars, manufacturers of Cl and Cl compounds, water purification plants, swimming pools, and incineration of plastics containing Cl.

In an investigation of an accidental Cl spill from a tank car in Ontario this writer observed that the area of injury comprised a swath of about 500 feet wide and one-half mile long. The edges of the injury area were sharply delineated with one side of bushes displaying severe injury while the other sides were unaffected. The symptoms of Cl injury were diverse, and ranged from terminal and marginal necrosis and chlorosis, to interveinal lesions occurring both bifacially and on upper leaf surfaces only. Eastern cottonwood and silver maple displayed interveinal necrotic lesions, whereas a Bartlett pear displayed cupped leaves with a distinct marginal necrosis. Defoliation was severe on walnut, willow, silver maple, privet, lilac, sycamore, and peach. Blossoms were killed on lilac bushes, and fruit drop occurred on apple, pear, and peach trees.

Chlorine injury is localized in effect, and not systemic. Observations of Cl injury in the field 6 weeks after a fumigation in Ontario showed injury on the older leaves, whereas leaves formed subsequent to the fumigation were green and healthy. Chemical analysis showed higher chloride contents in the older injured leaves.

Brennan *et al.* (1965) fumigated various plant species with Cl and found the symptoms to vary from foliage bleaching to white, tan or brown necrosis. In alfalfa and begonia the necrosis was marginal, whereas on tomato, tobacco, radish, and cucumber the necrotic lesions were scattered over the leaf blade. In corn the necrosis appeared in streaks between the veins, and in onion and pine the necrosis was terminal. Alfalfa and radish were injured by 0.1 ppm of Cl for 2 hr. This threshold dosage places Cl between fluorides and SO_2 in phytotoxicity. The same workers (Brennan *et al.*, 1966) found that 1.00 ppm Cl for 3 hr was required to injure pine needles.

In Ontario, this writer found the following plants to be sensitive to Cl injury: peony, lilac, apple, pear, peach, cottonwood, spruce, catalpa, sycamore, Chinese elm, and walnut. Intermediate in sensitivity were maple, willow, balsam fir, hydrangea, privet, and white ash. Resistant plant species included rose, chrysanthemum, cedar, juniper, coleus, mugho pine, red pine, hawthorne, and red oak.

In fumigation experiments the extent of Cl injury was reduced when plants were under conditions of water stress (Brennan et al., 1965). Benedict and Breen (1955) found that injury to weeds also was reduced in Cl fumigation experiments when the plants were grown under conditions of low soil moisture.

I. HYDROGEN CHLORIDE

Plant damage caused by HCl was first observed in Europe over 100 yr ago. In the LeBlanc soda process, NaCl was treated with H_2SO_4 to produce Na_2CO_3 with HCl released as a by-product. Plastics containing Cl when combusted release HCl gas. Injury and death to white ash and white oak trees in the vicinity of a manufacturer of $AlCl_3$ in Ontario was investigated by this writer. Emissions of $AlCl_3$ in contact with moisture in the air formed AlO and HCl gas, the latter being phytotoxic.

The symptoms of HCl injury on white ash leaves in Ontario were mainly in the form of marginal necrosis and interveinal necrotic lesions. An increase in chlorides was found in the injured leaves when compared to uninjured leaves collected at increasing distances from the source of HCl gas. A number of trees had been killed by the gas within 200 ft of the source of the HCl emissions. Haselhoff and Lindau (1903) reported that HCl fumes caused bleached lesions and necrotic margins on broad leaves, and tip burn on coniferous needles. Viburnum and larch seedlings were killed by exposures to 5 to 20 ppm of HCl gas for 2 days. Shriner and Lacasse (1969) injured tomato plants with 5 ppm of HCl for 2 hr. The injury on the leaves progressed through interveinal bronzing to necrosis. Godish and Lacasse (1969) exposed tomato plants to 8 to 10 ppm HCl gas for 2 hr and found symptom expression to be a function of relative humidity. No injury occurred at 40% relative humidity, with symptoms of injury progressing through undersurface glazing and interveinal bronzing to complete leaf collapse as relative humidity was increased to 65%.

Heck et al. (1970) listed plants sensitive and tolerant to HCl gas based on a limited number of studies. Sugar beet, cherry, larch, maple, tomato and viburnum were placed in the sensitive category, whereas beech, fir, maple, and pear were considered tolerant to HCl gas.

J. ETHYLENE

Except for ethylene the hydrocarbon group is not considered toxic to vegetation at concentrations found in ambient air. However, hydrocarbons along with NO are the primary ingredients in the photochemical reaction which result in secondary toxic smog products.

Ethylene has the potential to become a major phytotoxicant. It can injure vegetation at extremely low concentrations and is emitted from various sources such as automobiles and polyethylene manufacturers. Ethylene is produced metabolically within plant tissues and increases in concentration rapidly during maturation stages, for example in apple fruit ripening.

Ethylene causes epinasty (drooping), chlorosis, necrosis, and abscission of plant leaves (Heck et al., 1970). On rose plants young leaves develop epinasty and older leaves display interveinal chlorosis. Necrosis of sepals on orchid flowers is another common effect of ethylene.

The African marigold developed epinasty after a 24 hr exposure to 1 ppb of ethylene. Dry sepal in orchid occurred after a 6 hr exposure to 0.1 ppm ethylene.

Heck *et al.* (1970) listed plant species according to their sensitivity to ethylene. Sensitive plants included carnation, cotton, cucumber, marigold, orchid, peach, philodendron, privet, rose, sweet potato and tomato. Tolerant plants included beet, cabbage, clover, oats, onion, radish, and sorghum.

K. AMMONIA

Injury to vegetation from NH_3 occurs most frequently from accidental discharges from tank cars. Either a tank car (train or truck load) overturns or a pressure hose transferring anhydrous ammonia from one tank to another ruptures. This writer investigated the effects of NH_3 on vegetation in Ontario which originated from a ruptured hoseline. About 1000 pounds of 82% anhydrous ammonia were released and severe injury occurred in vegetation for a distance of about 1500 ft from the source. Several types of injury symptoms were observed on foliage, such as interveinal, marginal, terminal, and basal markings. Iris and gladiolus leaves displayed bleached injuries; apple and peony leaves had brown lesions; and lily-of-the-valley had black-colored areas. White birch displayed various-shaped orange-brown lesions depending on the areas of the leaves which had been exposed during the fumigation. A copper beech displayed atypical symptoms of air pollution injury. Instead of a normally-occurring marginal necrosis, the margins of the leaves were unaffected, whereas the central portions had been killed by the NH_3. Maple trees – silver, sugar, box elder, and Norway were resistant to the NH_3 fumes. Chemical analyses revealed 2000 ppm and more of NH_3 within the injured leaves, whereas about 100 to 200 ppm were found in healthy leaves outside the injury area.

In laboratory experiments (Thornton and Setterstrom, 1940) 40 ppm of NH_3 for 1 hr injured buckwheat, coleus, sunflower, and tomato leaves. Slight marginal injury was produced with 16.6 ppm NH_3 for 4 hr.

Heck *et al.* (1970) listed plant species sensitive and tolerant to NH_3. Mustard and sunflower were considered to be sensitive whereas dandelion, peach, and pigweed were placed in the tolerant class. In Ontario, this writer found that white birch and beech were extremely sensitive to NH_3, whereas various maple species were quite tolerant.

L. MERCURY

Mercury has become one of the new dangers of the 1970's. It is considered relatively harmless in its inorganic form but in the presence of microorganisms in lake and river bottom sediments, and in the bodies of birds, fish, and animals it is chemically transformed to an organic alkyl Hg compound which is extremely toxic. Millions of pounds of Hg are used annually by North American industries in the manufacture of

Cl, caustic soda, and other products. In agriculture organomercury compounds are used as fungicides to protect grain seed and as pesticides to be sprayed on crops. These compounds include phenylmercuric acetate and methylmercuric dicyandiamide. It is estimated that from 30 to 40 pounds of inorganic Hg are lost for every 100 tons of Cl manufactured, and that 3% of the lost Hg is emitted to the surrounding atmosphere (USDHEW, 1971).

Mercury is slightly volatile at ordinary temperatures, but with increase in temperature the rate of vaporization increases rapidly. The atmospheric concentration of Hg will approximately double for every $10\,°C$ increase (Stahl, 1969).

Boussingault (1867) reported injuring plants with vapors from Hg over 100 yr ago. Injury to plants from Hg vapor has been reported from experiments and from observations in confined areas in greenhouses. Dimond and Stoddard (1955) found that injury on roses arose from the use of a mercurial-fungicide paint used inside a greenhouse. A Hg detector with a lower limit of $10\ \mu g\ m^{-3}$ was used to analyze the greenhouse air and no detectable Hg vapors were recorded. Leaf analysis showed approximately 5 times more Hg in injured leaves than in leaves of plants which had not been exposed to Hg.

Plants have a tendency to accumulate high concentrations of Hg. When plants were enclosed in a glass case with soil which had been moistened with a solution of bichloride of mercury, 317 ppm Hg were found in Briarcliff rose leaves after 4 days, 1087 ppm in tomato leaves after 51 hr, and 3044 ppm in basal leaves of tobacco after 7 days (Zimmerman and Crocker, 1934). Floral parts were most sensitive, followed by the oldest leaves of plants. Rose flowers are bleached, and corollas absciss from the receptable without opening. Stamens are injured also, turning nearly black in half mature buds. Zimmerman and Crocker (1934) also reported injury on older peach leaves changing from interveinal chlorosis to browning. At higher temperatures injury was more severe and the whole leaf turned brown. Injury on a fern frond showed that the degree of injury varied with the age of the leaf tissue.

Heck et al. (1970) segregated 42 plant species into different degrees of sensitivity to Hg vapor. Bean, fern, hydrangea, privet, sunflower, and willow were classified as being sensitive. A large number of plants including azalea, forsythia, lily, peach, white pine, strawberry, tobacco, and tomato were considered intermediate, whereas aloe, cherry, holly, and ivy were listed as being tolerant.

M. PARTICULATE MATTER

Particulate matter may be solid or liquid with the size of the particles varying from a fraction of $1\ \mu$ in diam to over $100\ \mu$. Particulates which have an effect on vegetation include cement-kiln dust, magnesium-lime dust, carbon soot, H_2SO_4 aerosols, fluorides, Pb compounds, Ni compounds and salt spray.

Pierce (1910) found cement dust in combination with dew and light rain formed a thick crust on upper leaf surfaces of citrus trees, and interfered with the penetration of light required in photosynthesis with a resultant decreased starch formation. Anderson (1914) found reduced yield of cherries on the sides of trees facing a cement plant due to the prevention of pollen germination on dust-covered stigmas of flowers.

Darley (1966) applied cement-kiln dust to the expanding primary leaves of bean plants at the rate of 0.6 to 3.8 g m^{-2} for 8 hr for a period of 2 to 3 days and found CO_2 exchange reduced by 30%. Cement dust may cause chlorosis and death of leaf tissue by the combination of a thick crust deposit and alkaline toxicity produced in wet weather. Deciduous and coniferous trees are injured, with the latter occasionally killed.

This writer has observed in Ontario that cement dust deposited on apple leaves inhibited the action of hormonal sprays applied to the tree for the purposes of fruit retention and color development.

In the vicinity of a Mg producing industry in Ontario this writer found emissions of magnesium-lime dust to accumulate in the soil and raise the pH. When soils were excessively alkaline crop yields were reduced.

Soot-laden smoke emanating from a railroad roundhouse suppressed the growth of jack pine trees located in the immediate vicinity (Linzon, 1961). Injuries were related to the acidity of the soot particles and the tarry coating which clogged stomates and interfered with the normal exchange of gases.

Sulphuric acid droplets settling on dry leaves caused no injury, but in combination with moisture produced punctate spots on the upper surfaces of the leaves (Middleton et al., 1958).

Particulate fluorides are less harmful to vegetation than gaseous forms. McCune et al. (1965) found little or no injury produced in various plant species when exposed to cryolite (sodium aluminum fluoride dust). Accumulation of fluoride within plant tissues was much less than would occur from exposure to similar concentrations of gaseous fluorides.

Lead compounds emitted from automobiles, storage battery producers, and in the recovery of Pb from storage batteries have contaminated the soil and vegetation in the immediate vicinity. Lead is quite immobile and insoluble and there is little absorption of lead from the soil by crops (Schuck and Locke, 1970). Much of the particulate Pb deposited on vegetation is external and can be removed by simple washing.

Nickel compounds emitted from a Ni refinery in Ontario have been found by this writer to cause severe injury to vegetation. Several hundred ppm of Ni were found in foliage displaying both marginal and interveinal necrotic lesions.

De-icing compounds applied to highways in winter have resulted in injury to roadside trees caused by the splashing of the salt by passing vehicles. This writer investigated salt spray injury to roadside fruit trees in Ontario and found severe dieback on trees closest to the road accompanied by elevated chloride concentrations. The severity of injury on the trees decreased with the distance from the highway and became negligible usually beyond 150 ft. This distance was found to vary depending on frequency of salt treatment, highway traffic, height and density of roadside trees, and gradient of the highway.

3. Investigation and Prevention of Phytotoxic Effects

Air pollution injury to vegetation is steadily increasing in importance, and plant specialists must become more aware of pollution-incited diseases and find ways of controlling them. As with any plant disease, the pathologist investigating effects by air pollutants should consider the incitant, length of exposure, the injury symptoms, histopathology, environment, susceptibility, and control.

It has been evident from studies on the effects of phytotoxicants both in the field and in controlled fumigation experiments that plants vary considerably in their susceptibility. Some plant species and varieties possess such extreme sensitivity to air pollutants that they can be used as biological monitors of air pollution. Other plant varieties may be sufficiently resistant to withstand injury in the field, but may be injured in the laboratory by excessively high concentrations of pollutants.

Environmental factors which are conducive to optimum plant growth are usually the same factors which abet pollution damage. These factors include high light, temperature, relative humidity, and soil moisture. In addition, plant genotype, stage of growth, and tissue maturation determine the sensitivity of a particular species to pollution damage.

Investigations of alleged fume damage require knowledge of the results of pertinent research and details about the area under study. Visible injury, or growth suppression which may be caused by other agents such as insects, disease, adverse weather, poor nutrition, and mismanagement should be assessed. The concentration and duration of atmospheric pollutants must be measured and monitored, and accompanying meteorological parameters should be known. Injured vegetation should be taken to the laboratory for examination and assaying by pathological, histological, and chemical techniques.

Vegetation surveillance studies in areas of concern keep investigators informed of increasing or decreasing effects. Baseline studies can be conducted in agricultural or forested areas before a major pollution source becomes operational to determine the pre-pollution natural and endemic conditions. In the vicinity of pollution sources ecological studies consist of surveys of indigenous vegetation for the presence of chemical, insect, disease, and physiological injuries; the examination of established plots for the effects of air pollution on condition, growth, yield, and survival of vegetation; and the sampling and laboratory analyses of collected vegetation, soil, and pond water. Special studies can be carried out in the field using sensitive plant species and varieties raised in shelters which are equipped or not equipped with air filtering devices.

The significance of these studies is that if air sampling data correlated with vegetation data indicate that the biological component of the environment is in danger, then prompt abatement action is necessary.

The protection of plants from the adverse effects of phytotoxicants cannot be carried out in exactly the same manner as is possible with disease-causing organic reproductive bodies. A pollution-diseased plant cannot infect another plant, thus there is no need for a quarantine or for eradication of the affected plants. In certain instances, sprays and dusts have protected plants from pollution damage. The development of resistant varieties holds some promise. However, the best control is to reduce the concentrations of noxious pollutants at their sources so as not to exceed the accepted regional air quality standards for agriculture and forestry.

References

Adams, D. F., Mayhew, D. J., Gnagy, R. M., Rickey, E. P., Koppe, R. K., and Allen, I. W.: 1952, *Ind. Eng. Chem.* **44**, 1356.

Anderson, P. J.: 1914, *Plant World* **17**, 57.

Applegate, H. G. and Durrant, L. C.: 1969, *Environ. Sci. Tech.* **3**, 754.

Barrett, T. W. and Benedict, H. M.: 1970, Report No. 1., TR-7 Agric. Committee, APCA, Pittsburgh, Pennsylvania.

Benedict, H. M. and Breen, W. H.: 1955, Proc. Third National Air Poll. Symp., Pasadena, Calif.

Berry, C. R. and Ripperton, L. A.: 1963, *Phytopathology* **53**, 552.

Bleasdale, J. K. A.: 1959, *Symp. Inst. Biol.* **8**, 81.

Boussingault, M.: 1867, *Acad. Sci.* **64**, 924.

Brandt, C. S. and Heck, W. W.: 1968, *Air Pollution*, Academic Press, New York.

Brennan, E. G., Leone, I. A., and Daines, R. H.: 1950, *Plant Physiol.* **25**, 736.

Brennan, E. G., Leone, I. A., and Daines, R. H.: 1965, *Int. J. Air Water Pollution* **9**, 791.

Brennan, E. G., Leone, I. A., and Daines, R. H.: 1966, *Forest Sci.* **12**, 386.

Brisley, H. R., Davis, C. R., and Booth, J. A.: 1959, *Agron. J.* **51**, 77.

Costonis, A. C. and Sinclair, W. A.: 1969, *Phytopathology* **59**, 1566.

Darley, E. F.: 1966, *J. Air Pollution Control Assoc.* **16**, 145.

de Ong, E. R.: 1946, *Phytopathology* **36**, 469.

Dimond, A. E. and Stoddard, E. M.: 1955, *Conn. Agr. Expt. Sta. Bull.* **595**, 19.

Dochinger, L. S.: 1968, *J. Air Pollution Control Assoc.* **18**, 814.

Dochinger, L. S., Bender, F. W., Fox, F. L., and Heck, W. W.: 1970, *Nature* **225**, 476.

Dreisinger, B. R.: 1965, 58th Ann. Meeting Air Poll. Control Assoc. Toronto, Canada.

Drowley, W. B., Rayner, A. C., and Jephcott, C. M.: 1963, *Can. J. Plant Sci.* **43**, 547.

Dugger, Jr., W. M., Koukol, J., and Palmer, R. L.: 1966, 59th Ann. Meeting, Air Poll. Control Assoc., San Francisco, Calif.

Dugger, Jr., W. M., Mudd, J. B., and Koukol, J.: 1965, *Arch. Environ. Health* **10**, 195.

Godish, T. J. and Lacasse, N. L.: 1969, 62nd Ann. Meeting, APCA, New York City.

Guderian, R. and Strattman, H.: 1962, *Forschung Landis Nordrh* **118**, 7.

Haselhoff, E. and Lindau, G.: 1903, *Leipzig* **11**, 203.

Heck, W. W.: 1968, *Ann. Rev. Phytopath.* **6**, 165.

Heck, W. W. and Tingey, D. T.: 1970, Second Int. Clean Air Congress, Washington, D.C.

Heck, W. W., Daines, R. H., and Hindawi, I. J.: 1970, Informative Rept. No. 1., TR-7 Agricultural Committee, APCA, Pittsburgh, Pennsylvania.

Heggestad, H. E. and Middleton, J. T.: 1959, *Science* **129**, 208.

Hill, A. C. and Littlefield, N.: 1969, *Environ. Sci. Tech.* **3**, 52.

Hill, A. C., Heggestad, H. E., and Linzon, S. N.: 1970, Informative Rpt. No. 1, TR-7 Agricultural Committee, APCA, Pittsburgh, Pennsylvania.

Hill, A. C., Pack, M. R., Treshow, M., Downs, R. J., and Transtrum, L. G.: 1961, *Phytopathology* **51**, 356.

Hill, Jr., G. R. and Thomas, M. D.: 1933, *Plant Physiol.* **8**, 223.

Hitchcock, A. E., Zimmerman, P. W., and Coe, R. R.: 1962, *Contrib. Boyce Thompson Inst.* **21**, 303.

Hitchcock, A. E., Weinstein, L. H., McCune, D. C., and Jacobson, J. S.: 1964, *J. Air Pollution Control Assoc.* **14**, 503.

Houston, D. B. and Stairs, G. R.: 1970, at VII Int. Symp. on Fume Damage, Essen, Germany.

Jacobson, J. S., Weinstein, L. H., McCune, D. C., and Hitchcock, A. E.: 1966, *J. Air Pollution Control Assoc.* **16**, 412.

Katz, M.: 1949, *Ind. Eng. Chem.* **41**, 2450.

Katz, M. and Shore, V. C.: 1955, *J. Air Pollution Control Assoc.* **5**, 144.

Katz, M. *et al.*: 1939, Bulletin No. 815, National Res. Council, Ottawa, Canada.

Ledbetter, M. C., Zimmerman, P. W., and Hitchcock, A. E.: 1959, *Contrib. Boyce Thompson Inst.* **20**, 275.

Lee, T. T.: 1965, *Can. J. Botany* **43**, 677.

Leone, I. A., Brennan, E. G., and Daines, R. H.: 1966, *J. Air Pollution Control Assoc.* **16**, 191.

Linzon, S. N.: 1958, Joint Pub. Ontario Dept. Lands and Forests, Ontario Dept. of Mines, Toronto, Ontario.

Linzon, S. N.: 1961, *Forestry Chronicle* **37**, 102.

Linzon, S. N.: 1965, *Forestry Chronicle*, **41**, 245.

Linzon, S. N.: 1966, *J. Air Pollution Control Assoc.* **16**, 140.

Linzon, S. N.: 1967, *Can. J. Botany* **45**, 2047.

Linzon, S. N.: 1969, *Handbook of Effects Assessment*, C.A.E.S., Pennsylvania State Univ., University Park, Penn.

Linzon, S. N.: 1970, Second International Clean Air Congress, Washington, D.C.

Linzon, S. N.: 1971, *J. Air Pollution Control Assoc.* **21**, 81.

McCune, D. C.: 1969, American Petroleum Inst., New York, N.Y.

McCune, D. C., Hitchcock, A. E., and Weinstein, L. H.: 1966, *Contrib. Boyce Thompson Inst.* **23**, 295.

McCune, D. C., Hitchcock, A. E., Jacobson, J. S., and Weinstein, L. H.: 1965, *Contrib. Boyce Thompson Inst.* **23**, 1.

Macdowall, F. D. H.: 1965, *Can. J. Plant Sci.* **45**, 1.

MacIntyre, W. H.: 1949, *Ind. Eng. Chem.* **41**, 2466.

MacLean, D. C., Roark, O. F., Folkerts, G., and Schneider, R. E.: 1969, *Environ. Sci. Tech.* **3**, 1201.

Mayrhofer, J.: 1893, *Z. Pflanzenkrankh* **3**, 50.

Menser, H. A. and Heggestad, H. E.: 1966, *Science* **153**, 424.

Menser, H. A., Heggestad, H. E., and Street, O. E.: 1963, *Phytopathology* **53**, 1304.

Middleton, J. T., Darley, E. F., and Brewer, R. F.: 1958, *J. Air Pollution Control Assoc.* **8**, 9.

Middleton, J. T., Kendrick, Jr., J. B., and Schwalm, H. W.: 1950, *Plant Disease Reptr.* **34**, 245.

Miller, P. R., Parmeter, Jr., J. R., Fleck, B. H., and Martinez, C. W.: 1969, *J. Air Pollution Control Assoc.* **19**, 435.

Miller, P. R., Parmeter, Jr., J. R., Taylor, O. C., and Cardiff, E. A.: 1963, *Phytopathology* **53**, 1072.

O'Gara, P. J.: 1922, Unpublished data in files of American Smelting and Refining Company.

Pack, M. R.: 1966, *J. Air Pollution Control Assoc.* **16**, 541.

Pierce, G. J.: 1910, *Plant World* **13**, 283.

Schuck, A. E. and Locke, J. K.: 1970, *Environ. Sci. Tech.* **4**, 324.

Setterstrom, C., Zimmerman, P. W., and Crocker, W.: 1938, *Contrib. Boyce Thompson Inst.* **9**, 179.

Shriner, D. A. and Lacasse, N. L.: 1969, *Phytopathology* **59**, 402.

Shupe, J. L., Olsen, A. E., and Sharma, R. P.: 1970, Impact of Air Poll. on Vegetation Conf., Ontario Section, TR-7 Committee, APCA, Toronto, Ontario.

Solberg, R. A., Adams, D. F., and Ferchau, H. A.: 1955, Proc. 3rd National Air Poll. Symp., Pasadena, Calif.

Stahl, Q. R.: 1969, Litton Systems and National Air Poll. Control Adminstr., Publ. No. APTD 69-40.

State of California, Dept. of Public Health, 1966, Berkeley, Calif.

Stoklasa, J.: 1923, *Urban and Schwarzenberg*, Berlin, Germany.

Suttie, J. W.: 1969, *J. Air Pollution Control Assoc.* **19**, 239.

Swain, R. E.: 1949, *Ind. Eng. Chem.* **4**, 2384.

Swain, R. E. and Johnson, A. B.: 1936, *Ind. Eng. Chem.* **28**, 42.

Taylor, O. C.: 1968, *J. Occupational Med.* **10**, 485.

Taylor, O. C.: 1970, Impact of Air Poll. on Vegetation Conference, Ontario Section and TR-7 Agricul. Committee, APCA, Toronto, Ontario.

Taylor, O. C. and Eaton, F. M.: 1966, *Plant Physiol.* **41**, 132.

Taylor, O. C. and MacLean, D. C.: 1970, Informative Report No. 1., TR-7 Agricultural Committee, APCA, Pittsburgh, Pennsylvania.

Thomas, M. D.: 1958, *Agron. J.* **50**, 545.

Thomas, M. D.: 1961, *Air Pollution*, W.H.O., Monograph Series, 46, Columbia Univ. Press, New York.

Thomas, M. D. and Hendricks, R. H.: 1956, *Air Pollution Handbook*, McGraw-Hill Book Co., New York.
Thomas, M. D. and Hill, G. R.: 1937, *Plant Physiol.* **12**, 309.
Thomson, W. W., Dugger, Jr., W. M., and Palmer, R. L.: 1966, *Can. J. Botany* **44**, 1677.
Thornton, N. C. and Setterstrom, C.: 1940, *Contrib. Boyce Thompson Inst.* **11**, 343.
Todd, G. W.: 1958, *Plant Physiol.* **33**, 416.
Treshow, M. and Pack, M. R.: 1970, Informative Report No. I., TR-7 Agricultural Committee, APCA, Pittsburgh, Pennsylvania.
USDHEW: 1971, Air Poll. Control Office Publication No. AP-80.
Weinstein, L. H. and McCune, D. C.: 1970, Impact of Air Poll. on Vegetation Conf., Ontario Section and TR-7 Agricultural Committee, APCA, Toronto, Ontario.
Widtsoe, J. A.: 1903, Bulletin, No. 88, Expt. Sta. Agric. College of Utah, Logan, Utah.
Wislicenus, H.: 1901, *Z. Angew. Chem.* **28**, 689.
Zahn, R.: 1961, *Staub.* **23**, 343.
Zimmerman, P. W. and Crocker, W.: 1934, *Contrib. Boyce Thompson Inst.* **6**, 167.
Zimmerman, P. W. and Hitchcock, A. E.: 1956, *Contrib. Boyce Thompson Inst.* **18**, 263.

AIR QUALITY SURVEILLANCE

GEORGE B. MORGAN and GUNTIS OZOLINS

Division of Atmospheric Surveillance

1. Introduction

Air quality surveillance is defined as the systematic collection and evaluation of aerometric and related data, which include information on pollutant concentrations, sources, and emissions, and on certain meteorological parameters. Information on pollution levels in ambient air is obtained through the operation of networks of monitoring stations, which can encompass long-term nation wide or even global networks as well as intensive, short-term sampling studies of specific problem areas and situations. The development of source and emission inventories – through stack measurements, questionnaires, and engineering calculations – provides information on the types and amounts of pollutants emitted into the air and identifies the sources of pollution. The recording of wind flows, temperatures, and mixing depths provides information needed to define the diffusion characteristics of the atmosphere. In every case, the product of surveillance is usable information – so collected, structured, and presented that it is useful to the researcher, the control official, the industrialist, and the concerned citizen.

Present-day surveillance of the nation's air quality is a cooperative effort involving local, state, regional, and Federal air pollution control agencies. Almost every state and major city in the U.S. now operates an air quality surveillance program. These range in size and complexity of instrumentation employed from the use of very simple sampling devices to the operation of sophisticated monitoring networks involving automatic sampling-analyzing instrumentation.

The emphasis in this chapter is placed on that portion of surveillance dealing with the measurement of atmospheric concentrations of air pollutants. In the following sections we will delineate the specific needs for air quality data, describe the scope and requirements of monitoring activities, and discuss the chief components of ambient air quality surveillance-sampling-measurement networks, laboratory support, and data handling and analysis. Finally, we will consider briefly some trends and patterns of air quality, both in urban and nonurban areas of the United States.

2. Objectives of Surveillance

An adequate program for the surveillance of ambient air quality is requisite for the control of air pollution. Monitoring operations enable us to identify pollutants emitted to or present in the atmosphere and then to observe their concentrations, patterns, and trends. The data base derived from monitoring is required in the (a)

McCormac (ed.), Introduction to the Scientific Study of Atmospheric Pollution, 152–163. All Rights Reserved.
Copyright © 1971 by D. Reidel Publishing Company, Dordrecht-Holland.

assessment of pollutant effects on man and his environment, (b) study of pollutant interactions, patterns and trends, (c) establishment of ambient air quality standards, (d) development of abatement tactics and control regulations, (e) judging compliance with and/or progress made toward meeting ambient air quality standards, (f) direct enforcement of control regulations, (g) activation of emergency procedures to prevent air pollution episodes, and (h) guidance of future land use and transportation planning. In short, air pollution surveillance initially demonstrates the need for control of air pollution (by defining pollution levels), then guides the development of control measures (by showing the degree of control needed), and finally indicates how effective the controls are (by allowing comparison with established standards and/or by comparison of 'before and after' values).

3. Requirements for Monitoring Networks

Development of an air quality monitoring network requires decisions regarding the pollutants to be measured and entails considerations such as the number and types of stations needed, their location, choice of instrumentation, frequency of sampling, and the like. Clearly, these requirements are quite different for Federal programs and those of state and local agencies. The monitoring conducted by state and local air pollution control agencies is directed toward enforcement activities. These agencies use air quality data to appraise concentrations of specific pollutants, to determine if these concentrations exceed standards, to direct control actions, to determine ambient air quality throughout the metropolitan region, and to activate emergency control procedures to prevent air pollution episodes.

The Federal monitoring system, on the other hand, provides a uniform data base throughout the country against which all other air quality data can be verified; Federal surveillance systems measure pollutants that are expensive or unusually difficult to analyze; they identify and quantify new or newly recognized pollutants; provide data, field stations, laboratories, and personnel for research in measurement techniques; and demonstrate the impact of pollutant emissions on the air quality of both urban and nonurban areas.

The following discussion is directed toward the design of a monitoring network for metropolitan areas. It is within these regions that monitoring activities must be especially capable of supplying the requisite air quality data base to guide and direct enforcement activities.

A. POLLUTANTS

In carrying out the provisions of the current Federal legislation (Clean Air Act as amended, December 1970), the States must provide for adequate monitoring of the six air pollutants for which national ambient air quality standards have been promulgated. These are total suspended particulates, SO_2, CO, NO_2, photochemical oxidants, and total hydrocarbons. In addition, monitoring is generally conducted for certain other pollutants, depending on local conditions. These may include the various trace metals, odors, fluorides, Hg, and others. Since most of these are generally associated with specific sources, area-wide networks are not required; sampling is conducted near the sources, generally as special studies. Metropolitan area monitoring networks therefore are typically composed of instrumentation for measuring the six major air pollutants. These, with some exceptions, are common to all major metropolitan areas.

B. NETWORK SIZE

The question of how many stations are required to provide adequate surveillance of a region's air quality has plagued control agencies for some time. Because no rigorous mathematical methods, or any other methods, are available for determining the number and location of sampling sites, experience and judgement are essential.

The number of sampling stations required depends primarily on the existing pollution levels, their geographical and temporal variability, and the size of the region. The number of sampling stations must be adequate to allow definition of the area or areas where ambient concentrations may be expected to exceed those designated in air quality standards. Agencies must also gather information on air quality in other areas, including the nonurban portions of the Air Quality Control Regions.

Pollution levels and the size of the region can, to some degree, be related to the total population of the region. A first approximation of the number of stations required within a region therefore can be obtained from a graph relating population to network size (Figure 1). The curves in Figure 1 show a spread suggesting a minimum and a maximum number of stations for each population class, depending on the extent and degree of pollution. For example, a region of 1 million inhabitants with a severe SO_2 problem may require up to 25 samplers, whereas one of similar size with a minimal SO_2 problem would require 10 samplers. Although the curves should provide good estimates for application to population-related pollutants (usually from motor vehicles), such as CO, HC, NO_2, and oxidant, they do not necessarily apply as well to SO_2 and particulate matter. For the latter pollutants, industrial complexity and fuel-use patterns in the region strongly influence the pollution levels and thus affect network size regardless of the population.

Surveillance of SO_2 and NO_2 requires an additional decision concerning the mixture of mechanical and automatic samplers. A first approximation can again be obtained from Figure 1. The curve for mechanical samplers provides the estimate for the total number of SO_2 samplers needed. The difference between the estimates for mechanical and automatic samplers is the number of mechanical samplers required.

Figure 1 is intended only as a general guide to network size. The curves are based on a qualitative evaluation of cities of different population classes in terms of their existing networks, pollution patterns, geographic distribution of sources, and the like. The relationship between population and network size was derived from such investigations, combined with experience and knowledge. In general, population is a good index to network size, and population data are easily obtainable. In certain

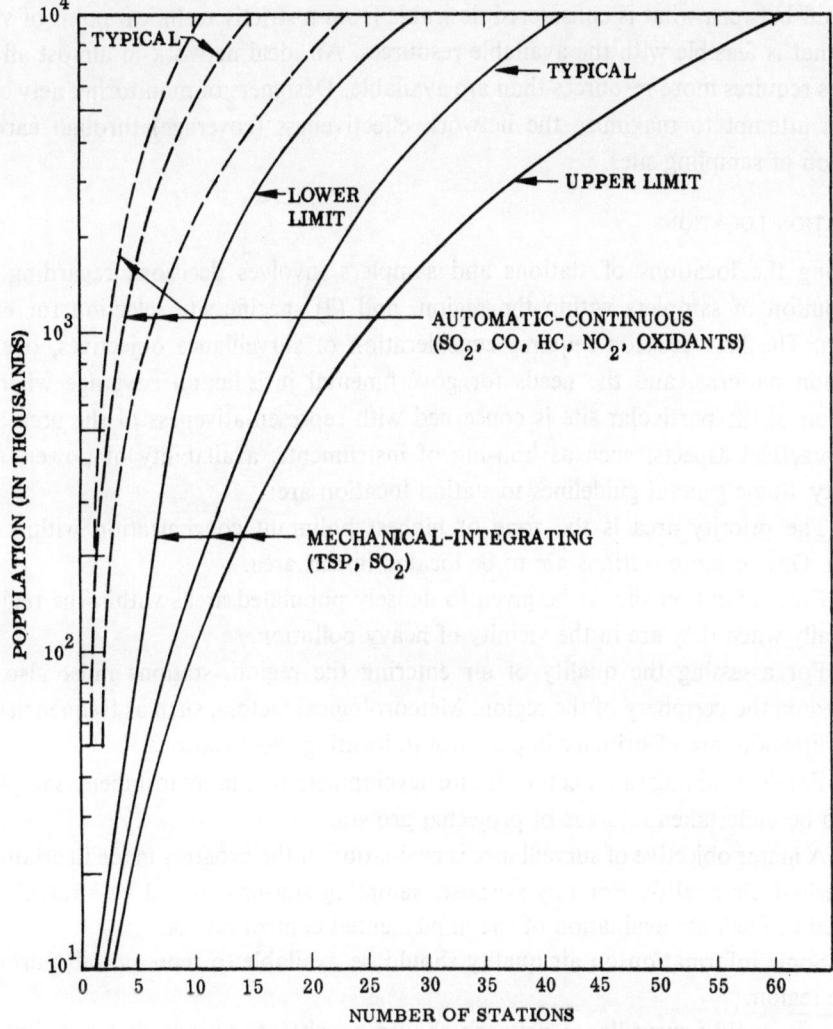

Fig. 1.   Number of stations versus region population.

situations, however, such as the relative absence of SO_2 pollution in the western portion of the United States, these curves are not applicable. In these situations, additional information, such as source strengths and their locations, is essential before network size can be determined.

Some other techniques for estimating network size have been developed and applied. One such method incorporates information on existing levels of pollution as a function of the area of the region. Whatever method is used to develop requirements of network size, it is likely that after more information on the region's air quality becomes available, the number of sampling stations and their location will need to be changed.

An important consideration in determining network size, of course, is the availability of resources. The selection of the monitoring system must by necessity involve a trade-off between what is considered desirable from a strictly technical point of view and what is feasible with the available resources. An ideal network in almost all instances requires more resources than are available. Designers of monitoring networks should attempt to maximize the network effectiveness (coverage) through careful selection of sampling sites.

C. STATION LOCATIONS

Selecting the locations of stations and samplers involves decisions regarding (1) distribution of samplers within the region, and (2) specific site selection for each station. The first decision requires consideration of surveillance objectives, overall pollution patterns, and the needs for governmental jurisdiction coverage whereas selection of the particular site is concerned with representativeness of the area and with practical aspects, such as housing of instruments, availability of power, and security. Some general guidelines to station location are:

(1) The priority area is the zone of highest pollutant concentration within the region. One or more stations are to be located in this area.

(2) Close attention should be given to densely populated areas within the region, especially when they are in the vicinity of heavy pollution.

(3) For assessing the quality of air entering the region, stations must also be situated on the periphery of the region. Meteorological factors, such as frequencies of wind direction, are of primary importance in locating these stations.

(4) For determining the effects of future development on the environment, sampling should be undertaken in areas of projected growth.

(5) A major objective of surveillance is evaluation of the progress made in attaining the desired air quality. For this purpose, sampling stations should be strategically situated to facilitate evaluation of the implemented control tactics.

(6) Some information on air quality should be available to represent all portions of the region.

The air quality surveillance network should consist of stations that are situated primarily to document the highest pollution levels in the region, to measure population exposure, to measure the pollution generated by specific classes of sources, and to record the nonurban levels of pollution. Many stations will be capable of meeting more than one of these criteria, e.g., a station located in a densely populated area could measure the population exposure and could also monitor the effectiveness of controls of emissions from domestic space heating, if such is part of the overall control strategy.

D. SAMPLING SITE CHARACTERISTICS

The preceding section gave guidelines for the general distribution of sampling stations within a region. Selection of a particular site for a single sampler or a complex station is equally important. It is essential that the sampler be situated to yield data representative of the location and not unduly influenced by the immediate surroundings. No definitive information is available concerning how air quality measurements are

TABLE I

Frequency of Sampling by Instrument Type Within Area

	Type of instrument	Areas above std.			Other urban areas				Nonurban areas			
		Con't.	Daily	Every third day	Con't.	Daily	Every third day	Every sixth day	Con't.	Daily	Every third day	Every sixth day
Particulates												
Total suspended particulate[a]	M[b]	×	×			×	×					×
Pb	M		×				×					
Polycyclic organic matter	M		×				×					
Fluorides	M		×				×					
Gases												
SO₂	M/A[c]	×	×		×		×				×	
CO	A	×	×		×		×					
Total HC	A	×[d]			×							
Non-methane HC	A	×	×		×							
NO₂	M/A	×	×		×		×				×	
NOₓ	M/A	×	×		×		×					
Oxidant (O₃)	M/A	×[d]	×		×							

a Spot-tape samplers provide an indication of TSP on a less than 24 h basis.
b M represents mechanized or integrated samplers.
c A represents automatic or continuous samplers.
d Where oxidant standards are based upon 3 h (6–9 a.m.) measurements of HC and NOₓ, the requisite data may be obtained from either continuous in-struments or mechanized samplers.

affected by the nearness of buildings, height from ground, and the like. We can, however provide guidelines that should be considered in the site selection:

(1) Uniformity in height above ground level is desirable for the entire network within the region. Some exceptions may include street canyons, high-rise apartments, and special-purpose samplers.

(2) Constraints to air flow from any direction should be avoided by placing inlet probes away from buildings (at least 3 m) or other obstructions. Inlet probes should be placed to avoid influence of convection currents.

(3) The surrounding area should be free from stacks, chimneys, or other local emission points.

(4) An elevation of 3 to 6 m is suggested as the most suitable for representative sampling, especially in residential areas. Placement above 3 m prevents most reentrainment of particulates, as well as the direct influence of automobile exhaust.

E. SAMPLING FREQUENCY

The sampling frequencies for mechanical samplers and the averaging times for automatic samplers are dictated by the ambient air quality standards. For example, if standards include concentration limits expressed in terms of days, hours, or minutes, then the sampler must have an averaging time appropriate for resolving days, hours, or minutes.

Although standards for total suspended particulates (TSP) and SO_2 are prescribed in terms of annual averages and maximum daily concentrations, it is impractical to operate the *entire* network on a daily basis. Adequate coverage may be maintained with intermittent sampling at frequencies calculated statistically to provide the desired levels of precision. Suggested sampling frequencies are presented in Table I, which related frequency of sampling to the degree of pollution, ranging from sampling every third day in highly polluted zones to every sixth day in nonurban zones.

4. Components of a Monitoring System

An air quality monitoring system is composed of three distinct but interrelated elements: (1) sampling-measurement networks, (2) laboratory support, and (3) data handling and analysis. With automatic (continuous) instrumentation the need for routine laboratory support is greatly reduced but a problem of data transmission, validation, and reduction is introduced.

A. INSTRUMENTATION

A variety of sampling devices and instruments is being used to collect samples and measure ambient air quality. Mechanical samplers are most generally used to collect integrated samples in the field. ('Integrated' in that sampling conducted over a given time period, say 24 h, yields a single sample to represent the entire time period.) The most common of these devices is the high-volume (Hi-Vol) sampler, which collects particulate pollutants on glass-fiber filters. Analysis of these samples provides information on concentrations of total suspended particulates, organic materials trace metals, and other organic and other inorganic pollutants. In addition to glass-fiber filters, membrane filters are used for collection of samples for subsequent analysis of pollutants such as Zn, asbestos, B, and silicates. Impactors of different designs are used to measure fractions of suspended particulates of various particle size ranges. Mechanical bubbler devices are used to collect SO_2, NO_2, oxidants, H_2S, aldehydes, CO_2, and NH_3. These samplers, although typically designed for the collection of 24 h integrated samples, can be modified to collect 1 or 2 h samples in sequence.

In automatic sampler-analyzers the collection and analytical processes are combined in a single device. This type of instrument produces a continuous analysis, with the output in a machine-readable format or in a form suitable for telemetry to a central data-acquisition facility. Continuous analyzers are now available for CO, CO_2, NO, NO_2, SO_2, total hydrocarbons, CH_4, total oxidants, O_3, and H_2S sulfide. Although a large number and variety of these analyzers are on the market, only a few have been properly field tested to determine their limitations (including interferences), reliability, and durability.

In the past, the operation of most mechanized and automatic analyzers has been based on wet chemical methods. These methods are not entirely satisfactory for typical field applications because of certain inherent problems: they must be attended frequently, reagents are unstable, and the instruments require complex plumbing and accurate solution pumps, which render them bulky and heavy. Future instruments will utilize the physical or physicochemical properties of pollutants for identification and quantification. A number of such new instruments are now in the final stages of development and field testing. Extensive research is also in progress to develop remote sensing devices to be used at ground level, in aircraft, and in earth satellites for area scanning and profiling.

B. LABORATORY ANALYSIS

Support of surveillance networks requires laboratory operations ranging from very simple to highly complex. The requirements for laboratory support, in terms of size and complexity, are dictated by the pollutants of interest in the region under surveillance, the degree of pollution, and the size of the networks. Most regions require equipment for analyses of total suspended particulates, SO_2, NO_2, and oxidants, and should provide for calibration of all collecting and measuring devices and preparation of reagents. Some regions require laboratory capability for analyses of trace elements, fluorides, and other pollutants. Large laboratories with requirements for multiple analyses are considering automated laboratory procedures. Applications of new laboratory techniques, such as microwave absorption spectrometry, electron or ion microprobe analysis, and neutron activation analysis, have been reported recently. In EPA laboratories, emission spectrometry, atomic absorption spectrometry, chromatography, and other techniques are now being computer-interfaced for routine use.

C. DATA HANDLING AND ANALYSIS

Processing and reporting the data acquired in surveillance constitute a vital part of any successful surveillance system. The data must be reported in a format that can be used conveniently in a variety of applications. On the Federal level, EPA plans to meet these requirements with a systematized information system, designated as the National Aerometric Data Information Service (NADIS). NADIS provides a means for air surveillance activities at all levels (Federal, State, regional, local) to participate in the National Aerometric Data Bank, in which data from across the nation are collected, stored, and retrieved to meet routine and emergency needs. NADIS provides for an orderly and timely flow of aerometric data into and out of the Data Bank.

Currently EPA's information service emphasizes making valid data available to all users as rapidly as possible. It is hoped, however, that more effort can be expended in analyzing and interpreting the data, in attempts to determine whether ambient concentrations are increasing or decreasing from year to year, nation wide and in specific locations, and what factors are affecting these changes.

Some beginnings have been made in this type of data interpretation, particularly in delineating patterns and trends in ambient air quality throughout the nation. Some of these trends are briefly discussed in the following section.

5. Trends in Urban Air Quality

A. PARTICULATES AND SO_2

Of all the identified pollutants, suspended particulates and SO_2 have been most extensively measured and studied. A multiple-station network for measuring these pollutants in some of the largest U.S. cities has been in operation for over 15 yr, and a considerable body of data is now available. The recorded levels show significant variations, ranging from extremely high concentrations in many cities to moderate and low levels in others. A nationwide overview of suspended particulate and SO_2 pollution is illustrated in Figures 2 and 3, which show the number of stations recording specified levels. As the illustrations indicate, a significant number of locations exceeds the levels now being established as air quality standards in the United States.

Combined temporal-spatial trends in levels of suspended particulate were analyzed on the basis of data from 20 nonurban and 58 urban sites over a period of 9 yr. A smoothed plot of all the data for the 20 nonurban sites indicated an unrelenting upward trend – toward higher levels of particulate pollution. In contrast, a similar analysis of data for the 58 urban sites for 10 yr shows a statistically significant decrease of about 7%. This decrease, indicating trends at single city-center sites is attributed mainly to control measures being applied to particulates, which are the most conspicuous form of air pollution. The downward trend also results from use of cleaner fuel associated with higher standards of living, and to a certain amount of decentralization of polluting sources, outward from the city core. It is important to realize that this downward trend is only for particulate matter. Conspicuous as this form of

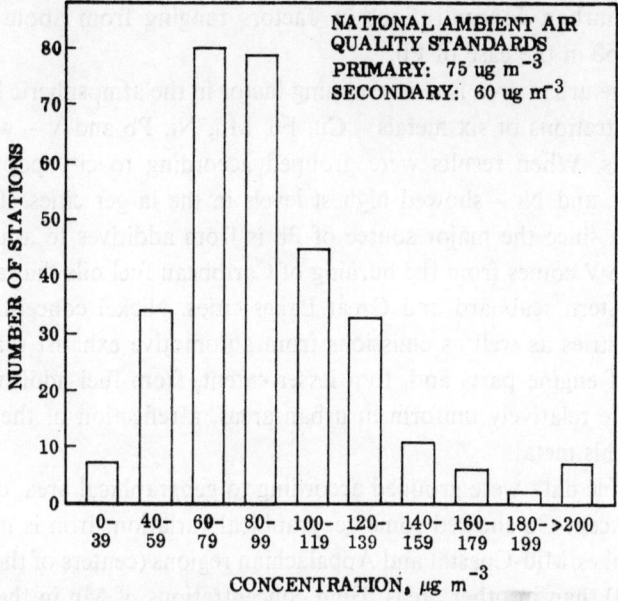

Fig. 2.   Distribution of annual station averages for suspended particulates in 1968.

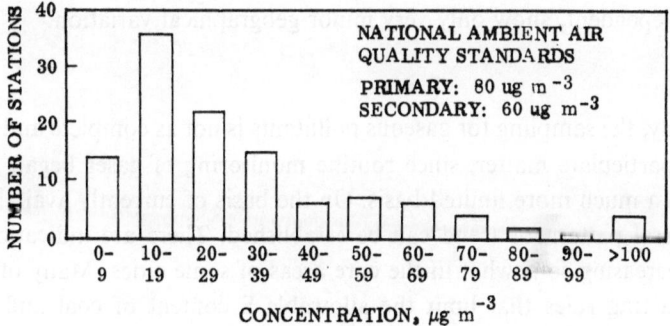

Fig. 3.   Distribution of annual station averages for SO₂ in 1968.

pollution is, particulate matter represents only 1 % of the total weight of the six major gaseous pollutants typically found in urban air – SO$_2$, CO, NO, NO$_2$, hydrocarbons, and oxidants.

B. METALS

Samples of total suspended particulates collected at various urban and nonurban sites have been analyzed for certain inorganic and organic substances that exhibit biological activity. These substances include sulfates, nitrates, fluorides, benzene-soluble organics, benzo(a)pyrene, Cu, Cr, Mn, Ni, Sn, Ti, V, Fe, Zn, and Pb. The relative concentrations of these metals in the atmosphere are quite different with Fe, Pb, and Zn being the most abundant. Urban metal concentrations are higher than the remote nonurban concentrations by factors ranging from about 2 for Cu to approximately 50 in the case of Pb.

The size of the urban area is a determining factor in the atmospheric levels of some metals. Concentrations of six metals – Cu, Fe, Mn, Ni, Pb and V – were measured in various cities. When results were grouped according to city population, three metals – Pb, V, and Ni – showed highest levels in the larger cities. This finding is understandable, since the major source of Pb is from additives to automotive fuels and most of the V comes from the burning of Caribbean fuel oils that are being used in the large eastern seaboard and Great Lakes cities. Nickel concentrations reflect the metal industries as well as emissions from automotive exhaust that result from normal wear of engine parts and, to a lesser extent, from fuel additives. Iron concentrations were relatively uniform in urban areas, a reflection of the large natural abundance of this metal.

When the same data were grouped according to geographical area, concentrations of all metals except Cu showed some geographical variation. Iron is more prevalent in the Great Lakes/Mid-Coastal and Appalachian regions (centers of the manufacture of Fe and steel) than in other areas. High concentrations of Mn in the Appalachian area result from the unusually high values found in the Kanawha Valley. Vanadium concentrations are high in the Great Lakes/Mid-Coastal region because of the use of Caribbean fuel oils of high V content. Levels of Pb and Ni, already shown to be population dependent, show only very minor geographical variation.

C. GASES

Unfortunately, the sampling for gaseous pollutants is not as complete and widespread as that for particulate matter, since routine monitoring of gases began many years later and on a much more limited basis. On the basis of currently available data, no definite general pattern or trend can be established. There are indications that the levels are decreasing somewhat in the core areas of some cities. Many of these cities are now enacting rules that limit the allowable S content of coal and of fuel oil. This limiting of S content, together with urban renewal which replaces buildings that often used heavy fuels for space heating, is the primary cause of a slight downward trend in SO_2 concentrations.

Data on carbon monoxide, which comes principally from motor vehicles, show a levelling off associated with near-saturation of our downtown streets by automobiles. Unfortunately we have no data on nonurban gaseous pollutant measurements with which to explore spatial trends as we did with particulates, but there is no reason to conclude that similar dispersion is not occurring. The principal problem may be not that pollution levels in city cores are intensifying, but that pollution is extending over larger and larger areas.

D. NEEDS FOR TREND ANALYSIS

Another urgent need in trend analysis concerns the evaluation of high pollution values that is important in relation to short term effects. The 'stagnation conditions' associated with high air pollution levels occur only on 5 to 10% of the days. It would be interesting to assess the effects on the occurrence and severity of these high pollution episodes of the numerous physical factors previously mentioned; such data would provide an additional insight into the effects of our changing social patterns on environmental air quality.

The lack of sufficient amounts of data from which to establish trends in air quality reflects the recentness of our society's attention to the impact of air pollution on our environment and also the high costs associated with collecting data of quality and quantity sufficient to allow assessment of trends.

What are the prospects that trends in our production of pollutants will be upward? Our national expectations are to double the population by 2025. Energy needs of the nation are now doubling every 10 yr, and this rate of increase is expected to continue for the next few decades. This, then, tells us that our potential to pollute will increase immensely – probably by a factor of 6 or 8. These increases will not be the same for all pollutants, of course. But most assuredly we will witness great increases in our abilities to produce pollutants such as CO, SO_2, NO_2, hydrocarbons and other organic compounds, fine particulates, odors, and others. Even though all of these pollutants will be increasingly controlled at the same time that we are benefiting from the activities that produce them, we will experience an ever-increasing need for improved abilities to assess pollution trends. Only by continuous surveillance and by interpretive analysis of timely data can we hope to maintain any hard-won improvement in environmental quality.

Pollutants of manmade origin have their principal sources in our urban centers, and control of such sources will, if need be, reflect back to man's activities in these urban complexes. The full technological development of the underdeveloped nations in the world would probably cause a 6-fold additional increase in the world's potential to pollute which calls for serious thought as to how to stem this tide, particularly with regard to possible world-wide climatic changes.

In the future, development of resources on both a national and international basis will challenge our ingenuity to develop ways to minimize adverse effects and to preserve environmental quality so that man can live in health and happiness in his environment. But whatever these effects, the alarming portent of the future underlines the importance of improving our nation's technical and analytical capabilities for air quality surveillance.

GLOSSARY

A	Angstrom (10^{-8} cm)
$AlCl_3$	Aluminum chloride
AlF_3	Aluminum fluoride
AlO	Aluminum oxide
Ar	Argon
B	Boron
Ba	Barium
Be	Beryllium
Brackets [ ]	Chemical symbol in brackets indicates concentration of that chemical per unit volume
C	Carbon
Ca	Calcium
CaF_2	Calcium fluoride
CH_4	Methane
C_NH_{2N+2}	Saturated hydrocarbon
CH_2O	Formaldehyde-hydrocarbon fuel byproduct
Cl	Chlorine
cm	centimeter
CO	Carbon monoxide
CO_2	Carbon dioxide
Cr	Chromium
Cu	Copper
eV	Electron volts (usually electron volts per molecule); 1 eV per molecule in a reaction is the same as 23 kilocalories per mole
F	Fluorine
Fe	Iron
FeO	Iron oxide
g	Gram; 454 grams = 1 pound
H	Hydrogen
HCl	Hydrogen chloride
HCN	Hydrogen cyanide
HF	Hydrogen fluoride

HNO_2	Nitrous acid
HNO_3	Nitric acid
H_2O	Water
H_2S	Hydrogen sulfide
H_2SO_4	Sulfuric acid
H_2SiF_6	Hydrofluorosilicic acid
Hg	Mercury
$HgCl_2$	Mercuric chloride
Hg_2Cl_2	Mercurous chloride
$h\nu$	Quantum of energy of light having frequency ν vibrations per second
I	Iodine
IR	Infrared radiation
K	Potassium
	Temperature in degrees Kelvin
kcal	Kilocalories per mole, 23 kilocalories per mole is equivalent to 1 electron volt per molecule
kg	Kilogram; 1 kilogram = 2.2 pounds
km	Kilometer; 1.6 kilometers = 1 mile
Kr	Krypton
m	Meter
M	In reactions it is any atom or molecule
mg	Milligram (10^{-3} gram)
Mg	Magnesium
mm	Millimeter; 1 millimeter = 10^{-3} meter
metric ton	1 metric ton = 10^3 kilograms
Mn	Manganese
N	Nitrogen
N_2	Nitrogen molecule
NH_3	Ammonia
NH_4OH	Ammonium hydroxide
$(NH_4)_2SO_4$	Ammonium sulphate
NO	Nitric oxide
NO_2	Nitrogen dioxide
NO_3	Nitrogen trioxide
NO_x	Nitrogen oxides
N_2O	Nitrous oxide
N_2O_4	Dimer of nitrogen dioxide
Na	Sodium
Na_3AlF_6	Cryolite
Na_2CO_3	Sodium carbonate

NaCl	Sodium chloride
NaF	Sodium fluoride
Ne	Neon
Ni	Nickel
O	Oxygen atom
O_2	Oxygen molecule
O_3	Ozone
OH	Hydroxyl
P	Phosphorus
Pb	Lead
PbO_2	Lead dioxide
PAN	Peroxyacetyl nitrate
pH	Negative $\log_{10}$ concentration of H^+ ions in solutions; hence, acidity is a low number
ppb	Parts per billion
pphm	Parts per hundred million
ppm	Parts per million
Ra	Radium
rem	Roentgen-equivalent-man
S	Sulfur
SO_2	Sulfur dioxide
SO_3	Sulfur trioxide
SO_x	Sulfur oxides
SiF_4	Silicon tetrafluoride
SiO_2	Silicon dioxide
Sn	Tin
Sr	Strontium
Ti	Titanium
U	Uranium
uv	Ultraviolet radiation
V	Vanadium
Xe	Xenon
Zn	Zinc
μm	Micron or micrometer (10^{-6} meter)
μg	Microgram (10^{-6} gram)

INDEX OF SUBJECTS

Aerosol 2–3, 10–11, 18–19, 27, 29, 32, 46–51, 104–105, 109–112
    mixing ratio 16–17
Air density 13–14
Air quality 152–163
Allergens 9, 41–42
Allergies 108–110
Ammonia 13, 34, 37, 145, 159
Anatomy 99–102
Animals 25, 108
Anthracosilicosis 118
Ar 13, 14
Asbestos 47, 119–120, 159
Asthma 110
Atmospheric constituents 13
Atmospheric pollutants
    concentration 20–22
    definition 2–3, 11
    diffusion 53–93
    equilibrium equations 20–22
    history 4–6
    human health 97–128
    man-made 24–51
    natural 24–51
    potential 87
    precipitation effect 81–83
    sinks 77–81
    sources 2–3, 24–30, 53–61
    surveillance 152–163
    transport 53–93
    types 24–30, 53–61
    vegetation effects 131–149
Atmospheric regions 15

Bacteria 25–27, 31, 34, 38
Be 120–121
Bronchitis 107, 117

Cardiorespiratory mortality 106–107
CH_4 9, 13, 40, 159
Chemical kinetics 21
Chemosphere 22–23
Chimney 62–63, 72, 75–78, 85
Cigarette smoking 126–128
Cl 143–144
CO 13, 26, 72–73, 115–116, 122, 153, 157–163
    chemistry 39–40
    source 25–29, 38–40

CO_2 13–14, 19, 26, 76, 83
    chemistry 38–40
    mixing ratio 16–17
    concentration 38–39
    source 25, 28–29
Combustion 25–27, 31, 38, 47
Condensation nuclei 10, 13
Coriolis force 69–71
Cyclones 69–71

Diffusion
    classical model 54
    Fickian 55–56
    gradient theory 55–56
    statistical 57–60
    temperature effects 64
Droplets 10
Dust 2

Ethylene 144–145
Extinction coefficient 18–19
Eye irritant 41, 123–124

Fluorides 140–143, 154, 157, 159
Forest fires 25, 27, 47
Fossil fuels 19, 25, 30, 38
Fumigation 66, 69

H 13
    radioactive 44–45
HCl 6, 144
HCN 11
He 13
Health, definition 97–99
Health effects
    allergies 109–110
    anthracosilicosis 118
    abestosis 119–120
    asthma 110
    Be 120–121
    bronchitis 107, 117
    cigarette smoking 126–128
    CO 115–116
    digestive 100
    eye irritant 123–124
    Hg 121
    hydrocarbons 116
    industrial hazards 111–112

infectious diseases 110–111
NO_x 113–114
O_3 112–113
oxidants 112–113
particles 101–102, 117–122
Pb 121
pneumoconioses 109, 118–120
respiratory 100–108, 112–116
skin 99–100
SO_x 102, 114–115
toxicology 103–105
Hg 6, 45, 145–146, 154
HNO_2 37
HNO_3 37
H_2O 20, 22–23, 26–30
concentrations 17–18
mixing ratio 16–17
H_2S 9, 26–27, 30–32, 122–126, 159
chemistry 31
H_2SO_4 10–11, 30–34
Hydrocarbons 9, 19, 23–24, 116, 153, 157
source 25–29, 40–41

Incineration 25, 28
Industrial exposure 111–112
Industrial processing 25, 28, 47
Infectious diseases 110–111
Inhalation 100–105, 109–116
Internal combustion engine 25, 27–28, 35, 39–40, 46
Inversion layer 11

Kr 13–14
radioactive 44–45

Lapse rate 64–71

Medicine, environmental 97
Metals 153–157, 161–162
heavy 45–46
sources 45–46
Meteorology 53–93
instruments 83–84
Meteors 25–29, 47–49
Methane 9, 13, 40, 159
Mixing height 87–90
Mixing ratio 16
Mesopause 15–17
Mesosphere 15

N 24, 26
N_2 13–14, 22, 24
Ne 14
NH_3 13, 34, 37, 145, 159
$(NH_4)_2 SO_4$ 30, 33–34, 37
Nitrogen compounds
source 25–28, 34–38

chemistry 34–38
NO 23–24, 34–37, 159–161
NO_2 13, 23, 32–37, 122, 153–163
NO_x 113–114, 139–140
N_2O 13, 34

O 22–24, 37
O_2 13–14, 19–20, 22–24, 36–37
O_3 2, 13–14, 22–24, 26, 29, 36–37, 112–113, 122, 135–138
concentration 16–17
mixing ratio 16–17
OH 22–23, 30
Odor 122–123, 154
Oxidants 23–24, 31, 36–37, 41, 112–113, 135, 153–159

PAN 41, 138–139
Particles 2, 10–13, 78, 81, 101–104, 117–122, 146, 153–161
meteoric 29
source 25–28, 46–51
sulfates 47–49
Pb 45–46, 121, 157, 161–162
Photochemical processes 14, 22–24
Photodissociation 22
Photoexcitation 24
Photolysis 22–23, 29
Photosynthesis 14, 31
Physiology 99–102
Plants 25
allergen source 41–42
aerosol source 47
Plumes 59–64, 68, 85–87
buoyant 62–64
Pneumoconioses 109, 118–120
Power generation 25, 28
Precipitation 81–83

Ra 44
Radicals 23
Radioactivity 5–6, 42–45
Respiratory 100–108, 112–116
Roentgen 42–43
Rn 44

Scattering, particles 18–19
Scavenging 78
Sea spray 25–27, 30, 47
Skin effects 99–100
Smog 10–12, 32–33, 38, 40–41, 46, 69, 124
Sulfur
sources 24–34
SO_2 9–11, 23–24, 26–27, 80, 92–93, 102, 125–126, 131–138, 153–163
chemistry 31–34
SO_3 11, 23

SO_x 114–115
Space heating 25–28
Sr 44
SST exhaust 28–30
Stoke's Law 50–51
Stratosphere 15, 28–30, 47–51
Surveillance
    CH_4 159
    CO 153, 157, 159, 161, 163
    fluorides 154, 157, 159
    gases 157
    Hg 154
    H_2S 159
    hydrocarbons 153, 157
    instrumentation 158
    laboratory analysis 159
    metals 153, 157, 161–162
    networks 153–155
    NH_3 159
    NO 159, 161
    NO_2 153, 154, 157, 159, 161, 163
    objectives 152–153
    odors 154
    oxidants 153, 154, 157, 159
    particulates 153, 157, 159–161
    Pb 157, 161–162
    sampling frequency 158
    SO_2 153–163
Synergism 98, 137–138

Temperature inversion 11, 15, 65–70
Temperature stratification 64–67, 88
Terpenes 27, 40–41, 47
Thermosphere 15–17
Toxicology 103–105

Transport 53–93
Tropopause 12, 15, 29–30
Troposphere 15–17, 31, 46–49

U 44
UV 29

Vegetation damage 131–149
    Cl 143–144
    ethylene 144–145
    fluorides 140–143
        symptomatology 140–142
        dosage levels 142
    HCl 144
    Hg 145–146
    NH_3 145
    NO_x 139–140
    O_3 135–138
        symptomatology 135–136
        dosage levels 136–137
    oxidants 135
    PAN 138–139
    particulates 146
    SO_2 131–135
        dosage levels 133
        symptomatology 132–133
    synergism 137–138
Visibility 11, 18–19, 46, 123
Volcanoes 9–12, 25–32, 38, 47–51

Wind action 25–27, 41
Winds 53–93
Wind turbulence 53–64, 71–77

Xe 13–14